NEW

D0724366

Cancer Biomarkers

The Promises and Challenges of
Improving Detection and Treatment

Committee on Developing Biomarker-Based Tools for Cancer
Screening, Diagnosis, and Treatment

Sharyl J. Nass and Harold L. Moses, *Editors*

INSTITUTE OF MEDICINE
OF THE NATIONAL ACADEMIES

THE NATIONAL ACADEMIES PRESS
Washington, D.C.
www.nap.edu

THE NATIONAL ACADEMIES PRESS   500 Fifth Street, N.W.   Washington, DC 20001

NOTICE: The project that is the subject of this report was approved by the Governing Board of the National Research Council, whose members are drawn from the councils of the National Academy of Sciences, the National Academy of Engineering, and the Institute of Medicine. The members of the committee responsible for the report were chosen for their special competences and with regard for appropriate balance.

This study was supported by Contract Nos. HHSH25056133, HHSN261200611002C, 200-2005-13434, HHSM-500-2005-00179P, HHSP23320042509XI, and 223-01-2460 between the National Academy of Sciences and the Department of Health and Human Services. Any opinions, findings, conclusions, or recommendations expressed in this publication are those of the author(s) and do not necessarily reflect the view of the organizations or agencies that provided support for this project.

International Standard Book Number 13: 978-0-309-10386-2 (Book)
International Standard Book Number 10: 0-309-10386-X (Book)
International Standard Book Number 13: 978-0-309-66711-1 (PDF)
International Standard Book Number 10: 0-309-66711-9 (PDF)
Library of Congress Control Number 2007921549

Additional copies of this report are available from the National Academies Press, 500 Fifth Street, N.W., Lockbox 285, Washington, DC 20055; (800) 624-6242 or (202) 334-3313 (in the Washington metropolitan area); Internet, http://www.nap.edu.

For more information about the Institute of Medicine, visit the IOM home page at: **www.iom.edu.**

The serpent has been a symbol of long life, healing, and knowledge among almost all cultures and religions since the beginning of recorded history. The serpent adopted as a logotype by the Institute of Medicine is a relief carving from ancient Greece, now held by the Staatliche Museen in Berlin.

"Knowing is not enough; we must apply.
Willing is not enough; we must do."
—Goethe

INSTITUTE OF MEDICINE
OF THE NATIONAL ACADEMIES

Advising the Nation. Improving Health.

THE NATIONAL ACADEMIES
Advisers to the Nation on Science, Engineering, and Medicine

The **National Academy of Sciences** is a private, nonprofit, self-perpetuating society of distinguished scholars engaged in scientific and engineering research, dedicated to the furtherance of science and technology and to their use for the general welfare. Upon the authority of the charter granted to it by the Congress in 1863, the Academy has a mandate that requires it to advise the federal government on scientific and technical matters. Dr. Ralph J. Cicerone is president of the National Academy of Sciences.

The **National Academy of Engineering** was established in 1964, under the charter of the National Academy of Sciences, as a parallel organization of outstanding engineers. It is autonomous in its administration and in the selection of its members, sharing with the National Academy of Sciences the responsibility for advising the federal government. The National Academy of Engineering also sponsors engineering programs aimed at meeting national needs, encourages education and research, and recognizes the superior achievements of engineers. Dr. Wm. A. Wulf is president of the National Academy of Engineering.

The **Institute of Medicine** was established in 1970 by the National Academy of Sciences to secure the services of eminent members of appropriate professions in the examination of policy matters pertaining to the health of the public. The Institute acts under the responsibility given to the National Academy of Sciences by its congressional charter to be an adviser to the federal government and, upon its own initiative, to identify issues of medical care, research, and education. Dr. Harvey V. Fineberg is president of the Institute of Medicine.

The **National Research Council** was organized by the National Academy of Sciences in 1916 to associate the broad community of science and technology with the Academy's purposes of furthering knowledge and advising the federal government. Functioning in accordance with general policies determined by the Academy, the Council has become the principal operating agency of both the National Academy of Sciences and the National Academy of Engineering in providing services to the government, the public, and the scientific and engineering communities. The Council is administered jointly by both Academies and the Institute of Medicine. Dr. Ralph J. Cicerone and Dr. Wm. A. Wulf are chair and vice chair, respectively, of the National Research Council.

www.national-academies.org

v

Reviewers

This report has been reviewed in draft form by individuals chosen for their diverse perspectives and technical expertise, in accordance with procedures approved by the NRC's Report Review Committee. The purpose of this independent review is to provide candid and critical comments that will assist the institution in making its published report as sound as possible and to ensure that the report meets institutional standards for objectivity, evidence, and responsiveness to the study charge. The review comments and draft manuscript remain confidential to protect the integrity of the deliberative process. We wish to thank the following individuals for their review of this report:

Gerard Anderson, PhD, Johns Hopkins Bloomberg School of Public Health
Stanley Hefta, PhD, Bristol-Myers Squibb
Hedvig Hricak, MD, PhD, Memorial Sloan-Kettering Cancer Center
Carolyn D. Jones, JD, MPH, AdvaMed
Lawrence A. Loeb, MD, PhD, University of Washington
Beverly S. Mitchell, MD, Stanford University Medical Center
Scott D. Patterson, PhD, Amgen, Inc.
Eric Schadt, PhD, Rosetta Inpharmatics, LLC

Although the reviewers listed above have provided many constructive comments and suggestions, they were not asked to endorse the conclusions or recommendations nor did they see the final draft of the report before its release. The review of this report was overseen by **Melvin Worth, MD,** Scholar-in-Residence at the Institute of Medicine, and **Gilbert S. Omenn, MD, PhD,** University of Michigan Medical School. Appointed by the National Research Council and the Institute of Medicine, they were responsible for making certain that an independent examination of this report was carried out in accordance with institutional procedures and that all review comments were carefully considered. Responsibility for the final content of this report rests entirely with the authoring committee and the institution.

Acknowledgments

The Committee is grateful to many individuals who provided valuable input and information for the study, either through formal presentations or through informal communications with study staff and Committee members. In addition to the speakers, moderators, and invited discussants at the IOM workshop on developing biomarkers, as noted in the appendix, contributors to the study include Peter Bach (Centers for Medicare and Medicaid Services), Carol Bigelow (Centers for Disease Control and Prevention), Ellen Feigal (The Critical Path Institute), Arthur Holden (Pharmaceutical Biomedical Research Consortium), Gail Javitt (Center for Genetics and Public Policy), Dan McGowan (SEMATECH Media Relations), Barbara Mittleman (NIH Office of Science Policy), Greg Raab (Raab and Associates, Inc.), Wolf Rogowski (Institute of Health Economics and Health Care Management), Todd Skaar and David Flockhart (Indiana University), Sudhir Srivastava and Donald Johnsey (NCI Cancer Biomarkers Research Group), Sean Tunis (Health Technology Center), Judith Wagner (IOM Scholar in Residence), Sidney Wolfe (Public Citizen's Health Research Group), and Janet Woodcock (Food and Drug Administration). In addition, Margie Patlak assisted the committee by preparing some written background material on FDA oversight, CLIA, and the evaluation and adoption of clinical diagnostics for the report.

Contents

Boxes, Figure, and Tables

BOXES

FIGURE

TABLES

Summary

Biomedical scientists have long sought to identify ways to diagnose cancers at an early, curable stage or to select the optimal therapy for individual patients. Many cancer patients are diagnosed at a stage in which the cancer is too far advanced to be cured, and most cancer treatments are effective in only a minority of patients undergoing therapy. Thus, there is tremendous opportunity to improve the outcome for people with cancer by enhancing detection and treatment approaches. Biomarkers will be instrumental in making that transition.

Because of the heterogeneity among diseases and patients, recharacterization of disease in pathophysiological terms via biomarkers is key to the future of medicine. A biomarker is defined as any characteristic that can be objectively measured and evaluated as an indicator of normal biological processes, pathogenic processes, or pharmacological response to a therapeutic intervention. These indicators could include a broad range of biochemical entities, such as nucleic acids, proteins, sugars, lipids, and small metabolites, as well as whole cells or biophysical characteristics of tissues. Detection of biomarkers, either individually or as larger sets or patterns, can be accomplished by a wide variety of methods, ranging from biochemical analysis of blood or tissue samples to biomedical imaging. The primary focus of this report is in vitro diagnostic[1] tests. Although many of the challenges in bio-

---

[1]In this report, "diagnostic" is often used synonymously with "biomarker test." These terms refer to any laboratory-based test that can be used in drug discovery and development as well as in patient care and clinical decision making.

marker development are relevant to both biomedical imaging and in vitro diagnostics, in vivo imaging also entails a set of unique considerations, in part because it often requires administration of chemical agents, and thus has some similarity with drug development. Biomedical imaging will be addressed in a forthcoming workshop of the National Cancer Policy Forum, as this topic was beyond the scope of this report.

In recent decades, knowledge about the basic biology and biochemical pathways underlying cancers has increased tremendously, but translation of that knowledge to more effective patient care and better outcomes remains a challenge. Recent technological advances that enable examination of many potential biomarkers have fueled renewed interest in and optimism for developing biomarkers, and it is widely believed that biomarkers can and will be used to improve cancer screening and detection, to improve the drug development process, and to enhance the effectiveness and safety

---

BOX S-1   Summary of Recommendations to Develop Biomarker-Based Tools for Cancer

Methods, Tools, and Resources Needed to Discover and Develop Biomarkers (Chapter 2)

1. Federal agencies should develop an organized, comprehensive approach to biomarker discovery, and foster development of novel technologies.

2. Industry and other funders should establish international consortia to generate and share precompetitive data on the validation and qualification of biomarkers.

3. Funders should place a major emphasis on developing quantitative pathway biomarkers to broaden applicability.

4. Funders should sponsor demonstration projects to develop biomarkers that can predict efficacy and safety in patients for drugs already on the market.

5. Government agencies and other funders should sustain support for high-quality biorepositories of prospectively collected samples.

Guidelines, Standards, Oversight, and Incentives Needed for Biomarker Development (Chapter 3)

6. Government agencies and other stakeholders should develop a transparent process to create well-defined consensus standards and

of cancer care by allowing physicians to tailor treatment for individual patients—an approach known as personalized medicine. Some promising strides have been made in classifying tumors at the molecular level and in selecting patients who are more likely to respond to some targeted therapies. However, progress overall has been slow, despite considerable effort and investment, and there are still many challenges and obstacles to overcome before this paradigm shift in oncology can become a reality.

The committee was asked to examine questions regarding the discovery, development, adoption, and use of biomarkers for cancer screening, diagnosis, and treatment, with the goal of identifying obstacles to progress that could potentially be overcome through policy changes. The committee identified a number of challenges in biomarker research, development, and implementation and proposed 12 recommendations to foster progress in the field, as outlined in Box S-1. These recommendations fall into three general

guidelines for biomarker development, validation, qualification, and use.

7. The Food and Drug Administration and industry should work together to facilitate the codevelopment and approval of diagnostic-therapeutic combinations.

8. The Food and Drug Administration should clearly delineate and standardize its oversight of biomarker tests used in clinical decision making.

9. The Centers for Medicare & Medicaid Services should develop a specialty area for molecular diagnostics under the Clinical Laboratory Improvement Amendments.

Methods and Processes Needed for Clinical Evaluation and Adoption (Chapter 4)

10. The Centers for Medicare and Medicaid Services should revise and modernize its coding and pricing system for diagnostic tests.

11. The Centers for Medicare and Medicaid Services, as well as other payors, should develop criteria for conditional coverage of new biomarker tests.

12. As a component of conditional coverage, establish procedures for high-quality population-based assessments of efficacy and cost-effectiveness of biomarker tests.

categories: (1) methods, tools, and resources needed to discover and develop biomarkers; (2) guidelines, standards, and oversight needed for biomarker development; and (3) methods and processes needed for clinical evaluation and adoption. Although this report is focused on biomarkers for cancers, implementing these recommendations could have a broad positive effect on the development of biomarkers in general, thereby aiding progress in other areas of biomedical research and reducing the burden of other diseases as well. A great deal of work remains to be done, and keeping in mind the opportunity cost of investing in different areas of biomedical research, the committee's recommendations aim to streamline the biomarker discovery and development process, to make effective use of the available resources, and to develop a pathway for success that balances the need to encourage innovation while also ensuring that adequate standards for validation and qualification are met.

METHODS, TOOLS, AND RESOURCES NEEDED TO DISCOVER AND DEVELOP BIOMARKERS

Despite some notable achievements, only a few biomarkers are routinely used in oncology. Although advances in technology have made it easier to examine many potential biomarkers in a single experiment, discovery efforts are still hampered by the limitations of current technology. In addition, most candidate biomarkers never advance beyond the discovery phase, and the number of biomarkers validated for use in drug development or qualified for clinical applications is still very small. Obstacles to progress could be overcome or minimized by developing different strategies to foster the work, from discovery through development, as outlined in the first five recommendations. These approaches could lead to better biomarkers for the entire spectrum of cancer health care, from chemoprevention, early detection, and disease classification to drug development and treatment planning and monitoring.

Recommendation 1:
Federal agencies including, but not limited to, the National Institutes of Health (NIH) and the National Cancer Institute (NCI), the Food and Drug Administration (FDA), and the National Institute of Standards and Technology (NIST) should take a more organized, comprehensive approach to the

discovery of putative cancer biomarkers and the development of novel technologies.

- A highly directed, contract-based program could be effective in supporting the development of innovative biomarker discovery technologies.
- Extramural experts should be involved in all aspects of program planning, execution, and oversight.
- Successful implementation of this endeavor will require one federal agency to take responsibility for coordinating and overseeing the process.

Rationale

Biomarker discovery efforts to date have been piecemeal and unorganized. In addition, most current biomarker tests use technology that has been available for decades, and the ability to discover and develop new biomarkers is limited by the sensitivity, specificity, and capacity of current technology. Thus, there is a significant need for a more thorough and organized approach to discovery, as well as for new and improved technologies for biomarker discovery, particularly in the field of proteomics, which is more complex and has lagged behind advances in methods for analyzing nucleic acids. Such new technologies also will yield dividends in improved capabilities for understanding fundamental cellular processes in cancers and systems biology in general.

The NIH peer review process generally tends not to favor high-risk projects, and neither broad discovery efforts nor technology development has traditionally been a primary focus of NIH funding. However, that has been slowly changing in recent years, with several directed discovery projects and some recent initiatives by NIH that include the goal of improving technologies for protein detection and characterization. An organized, large-scale approach with a highly directed contract-based program would foster biomarker discovery and technology innovation and also provide incentives to academic researchers to undertake the work, since the career structure and reward system in academia is generally not conducive to technology development or large-scale discovery efforts. The involvement of multiple federal agencies is important, as no single agency is likely to have the needed expertise to address all issues, but it will be important for one agency to take the lead in organizing intra-institutional efforts. Given NCI's current funding level and recent initiatives and interest in biomarker discovery and development, it may seem an obvious choice for the lead agency for this endeavor, but to date it has not yet developed an adequate overarching leadership strategy.

Recommendation 2:
Industry and other funders of biomedical research should establish international public–private consortia, modeled after the SNP (Single Nucleotide Polymorphism) Consortium (see Box 2-5), to generate and share methods and precompetitive data on the validation and qualification of cancer biomarkers for specific uses.

Rationale

Collaborative, precompetitive projects (most likely unrelated to a particular drug) would be enabling to the field. The costs, uncertainties, and risks of developing biomarkers historically have made this work unappealing to pharmaceutical companies and diagnostic companies alike. Although industry perspectives are slowly changing regarding the strategic value of biomarkers, current business models are neither very viable nor attractive to investors. Furthermore, the quantity of data currently generated by any single company is likely to be inadequate for developing and validating biomarkers, and much duplicative work could be going on without ever being reported, so the process could be greatly improved and streamlined by sharing data and information that is already being generated.

Private companies normally are inclined to protect their data to maintain a competitive edge. However, the willingness of multi-national drug companies to share information via the SNP Consortium to achieve a common goal shows the feasibility of a collaborative approach for fostering precompetitive work that could benefit the entire field. Successful examples of sharing precompetitive data exist outside of biomedical science as well, including SEMATECH (Semiconductor Manufacturing Technology), which helped U.S. semiconductor suppliers develop new production tools and establish industry-wide consensus on product specifications. There is also precedent in the validation of a biomarker; the HIV Surrogate Markers Collaborative Group confirmed the usefulness of HIV RNA as a surrogate marker for testing new anti-HIV drugs. Given the past accomplishments of consortia, a number of new initiatives have recently been planned or launched with some concentration on biomarker validation, but most do not focus on cancer. One exception is the Oncology Biomarker Qualification Initiative, jointly funded by NCI, FDA, and CMS, which currently has only one project under way. Many more areas of cancer biomarker research could benefit from international public–private collaborations.

Recommendation 3:
Funders should place a major emphasis on research to develop quantifiable biomarkers of cell signaling pathways that will have the broadest applicability (e.g., across different tumor types and drugs, as well as other diseases).

Rationale

This approach could reduce the risk inherent in the biomarker development process by increasing their applicability. Biomarkers that are exclusively focused on a particular drug could become obsolete if the drug fails to gain FDA approval, or if therapy guidelines change. In contrast, markers that can identify biochemical pathways that are altered in cancers are more likely to be applicable to the development of any new drug that targets an essential pathway. For example, signaling pathway biomarkers that are validated in common cancers potentially could also be useful in rarer forms of cancer that are more difficult to study and would offer smaller returns to developers. Furthermore, pathway biomarkers could also be useful for early detection of multiple cancer types. Pathway biomarkers would also allow for a "systems" approach to diagnosis, treatment, and surveillance, recognizing that signaling pathways operate in the context of interconnected networks.

Recommendation 4:
Federal agencies and other funders should sponsor adequately powered demonstration projects focused on a single disease or pathway to discover and develop biomarkers that can predict safety and efficacy in individual patients (and thus select appropriate target populations) for drugs already on the market, as proof of principle and to establish a paradigm for biomarker development.

Rationale

The purpose of such demonstration projects would be to identify patient populations likely to respond to a drug, those likely to have resistance to a drug, and those likely to experience adverse reactions to a drug. A high-impact finding would not only improve treatment outcomes for patients, but could also define the biomarker field and catalyze the diagnostics and pharmaceutical industries and academia to undertake such studies for many cancers and therapeutics by establishing a viable route to market and by delineating a viable business strategy. Questions about how best to conduct such studies will need to be addressed early on. In particular, the studies must be well designed and adequately powered, but since patients are already taking the drugs, it should not be difficult to accrue participants for a study.

Recommendation 5:
NIH, NCI, and other funders should initiate and sustain funding for high-quality and highly accessible biorepositories of patient samples prospectively collected in conjunction with large cohort studies and clinical trials, and use of these prospectively collected samples should be encouraged for validating biomarkers. NCI should actively encourage and facilitate interaction among all interested biomarker developers and groups involved in clinical research, including therapeutic, screening, prevention, and cohort studies, to enable the prospective collection of high-quality patient samples that are intended to test specific hypotheses. The following would be important to ensure the quality and value of repositories:

- Providing sufficient funding to cover all essential biorepository components and activities, including:
    — Involvement of pathologists to assess sample quality and confirm diagnosis
    — Optimized sample collection and preparation
    — Capturing and consistently annotating clinical patient records
    — Medical informatics and database management
    — General administrative and maintenance costs.
- Adhering to standard operating procedures.

- Developing consensus on common data elements.
- Developing strategies for prioritizing and maximizing access to samples (including procedures for handling intellectual property issues).
- Developing strategies to ensure patient rights and privacy without impeding research, such as:
    — Reassessing the Privacy Rule established under the Health Insurance Portability and Accountability Act and promoting uniformity across states and institutions
    — Promoting interagency harmonization on informed consent to maximize the use and value of collected samples
    — Ensuring broad representation of extramural experts on oversight committees.
- Supporting biorepositories through public–private consortia in the long term, as proposed in Recommendation 2.

Rationale

Tumor samples have been collected and stored in repositories for many years, but there is substantial variability in methods, data elements, and quality. In addition, access to samples may be highly restrictive, and there is no central clearinghouse in which researchers can search for or obtain access to samples. When clinical trials end or funding for cohort studies is not renewed, the ability to maintain the biorepositories created in conjunction with the study is often lost. The samples collected in these prospective studies may be highly valuable for biomarker research and development, and NIH should consider continued funding for biorepository maintenance, even if the original study itself is not continued.

There are numerous examples of ongoing activities that could be instructive in this undertaking. For instance, the Multiple Myeloma Research Consortium Tissue and Data Banks provide an excellent model for the collection and use of patient samples for cancer research. Several NCI initiatives are also noteworthy, including the Early Detection Research Network's Informatics Infrastructure, which provides a model for defining common data elements and sharing information among researchers and institutions. The new Office of Biorepositories and Biospecimen Research is developing guidelines to optimize and unify operational, legal, and ethical policies and procedures for NCI-supported biorepositories, and it has launched a pilot study to test implementation of those guidelines.

GUIDELINES, STANDARDS, OVERSIGHT, AND INCENTIVES NEEDED FOR BIOMARKER DEVELOPMENT

The discovery of putative biomarkers is often reported in scientific journals, but validation of those potential markers for specific uses requires a great deal of additional investigation and study, which is often not undertaken because of the cost and complexity of the work. When these studies are undertaken, the standards used to demonstrate validity vary considerably, in part because there is no overarching leadership in the field of biomarkers to set uniform, consensus standards for biomarker development. The FDA and CMS have some authority over diagnostic tests, but oversight has been variable and unpredictable, and in many cases inadequate to ensure the safety, effectiveness, and value of tests on the market. Oversight by federal agencies has been evolving recently, with greater scrutiny of some tests by the FDA, and with CMS taking a greater interest in diagnostic tests. The FDA in particular has taken initial positive steps with the recent development of draft guidance documents. However, there is still a need for clarification, uniformity, and leadership in this area. The next four recommendations strive to improve the process of biomarker development and evaluation by making it more transparent, consistent, and effective.

Recommendation 6:

Government agencies (e.g., NIH, FDA, CMS, NIST) and non-government stakeholders (e.g., academia, the pharmaceutical and diagnostics industry, and health care payors) should work together to develop a transparent process for creating well-defined consensus standards and guidelines for biomarker development, validation, qualification, and use to reduce the uncertainty in the process of development and adoption. The appropriate federal agency should take responsibility to provide a leadership role in the process, coordinating and overseeing interagency activities.

- Different sets of guidelines will probably need to be developed for different applications, including the various stages of drug development and different types of clinical applications (e.g., prevention, screening, diagnosis, treatment planning, response monitoring, and surrogate endpoints), for different technologies, and for single biomarkers versus biomarker panels or patterns.

- A dynamic process will be needed so that guidelines can be revised as technologies or evidence change.
- Federal funding should stipulate adherence to publication guidelines.
- FDA, CMS, and industry should work together to develop guidelines for clinical study designs that will enable sponsors to run a single study (or a minimal number of studies) to generate adequate clinical data for review by both agencies.
- Postmarket surveillance will be needed to ensure quality and accuracy.

Rationale

Oversight, strategy, and ownership of the biomarker development process are key to success, but no federal agency currently takes responsibility for ensuring the clinical validity or utility of biomarkers. NIST has had a limited role in the biomedical sciences to date, but it has the appropriate experience to play a broader role in the establishment of standards for biomarkers, given adequate funding.

Professional organizations and other groups have developed numerous guidelines for publication of data or for clinical use of biomarkers; however, most are nonbinding. This piecemeal approach has created a patchwork of standards, with many gaps as well as some overlap, which can lead to conflict and confusion. Uniform consensus guidelines that cover the entire continuum are needed.

Currently developers often determine their own standards each time they consider a new biomarker, and competition to reach the market quickly creates an incentive to lower those standards. Variability in evidence standards applied by the FDA, CMS, and other health care payors can have a major impact on the cost of development and product revenue, so uniform standards would reduce the risk of development and make it easier to predict return on investment. Optimizing clinical study design will shorten the time to market, reduce cost and risk, and strengthen the evidence base for evaluation. It is important to strike a balance between fostering innovation and ensuring the validity and usefulness of tests.

Improved postmarket surveillance will be important to maintain high-quality standards because CMS oversight via the Clinical Laboratory Improvement Amendments (CLIA) appears not to be sufficient to ensure the accuracy of current biomarker tests. Accuracy problems with well-established clinical tests (e.g., immunohistochemistry for HER2 and estrogen receptor) underscore the need for greater postmarket oversight.

Recommendation 7:
The FDA and industry should work together to facilitate the codevelopment of diagnostic-therapeutic combinations.

- The FDA should more clearly delineate the expectations and requirements for diagnostic-therapeutic combination approval, and approval of the therapeutic and diagnostic should be linked, such that one is contingent on the other.
- Companies need to better integrate basic and clinical research, and emphasize the search for patient subpopulations based on theoretical and empirical evidence prior to phase III trials.
- Because more than one FDA center will be involved in the approval/clearance decisions for diagnostic-therapeutic combinations, the agency should clarify the roles of each center and focus on ensuring coordination among the centers to facilitate the review and approval process.
- The FDA should develop more dynamic ways of changing drug labels when new data for selecting appropriate target populations emerge.

Rationale

Coordinated development of diagnostics and therapeutics could help companies choose the most promising drug leads, optimize clinical trial designs, and facilitate rapid and effective adoption into clinical practice. However, diagnostics and therapeutics are currently developed separately, often by different entities. Timing is key for corelease and marketing of a diagnostic linked to a drug, but often there is a rush near the end of drug development to develop the diagnostic. As a result, the diagnostic may not be scrutinized as thoroughly. New strategies, methods, and infrastructure are needed to leverage and integrate the available data to better inform the biology.

The cost and risks of diagnostic development and validation are great when clinical validity and utility must be established, and they add substantially to the existing high cost of drug development. Companies may be unwilling to invest in diagnostic development in the earlier phases of testing when approval of the drug is so uncertain, because without drug approval, there may be no market for the diagnostic. Thus, devising strategies to share and minimize the costs and risks of codevelopment would foster this work.

Recommendation 8:
The FDA should clarify its authority over biomarker tests linked to clinical decision making and then establish and consistently apply clear guidelines for the oversight of those tests. In addition, the appropriate federal agency (e.g., the FDA or the Federal Trade Commission) should monitor and enforce marketing claims made about molecular diagnostics.

- A coherent strategy is needed to define and clarify the rules and their enforcement to make the process more transparent, to remove inconsistency and uncertainty, and to elevate standards and oversight.
- The FDA needs a dynamic process for updatating regulations to adapt to rapid changes in technology.
- The FDA needs additional resources if it is to make meaningful changes in this field.

Rationale

The FDA previously has claimed legal authority to assert jurisdiction over diagnostic tests, but generally it has chosen to limit oversight, most likely due to resource constraints. Recently, the FDA appears to be trying to create clarification and precedent on a case-by-case basis regarding molecular diagnostics, through warning letters and untitled letters and via nonbinding guidance documents. However, variability and unpredictability in FDA oversight can reduce interest and investment in developing innovative diagnostics. Moreover, inadequate evaluation and oversight could lead to harm for patients and unnecessary cost burden for society.

The Federal Trade Commission prohibits false or misleading advertising and has claimed it will take action against such advertising of genetic tests. But the agency's limited resources appear to be preventing it from following through on its commitment. To date, it has not exercised authority to enforce accurate marketing claims for molecular diagnostic tests.

Recommendation 9:
CMS should develop a specialty area for molecular diagnostics under CLIA.

Rationale

The minimum generic standards set by CMS under CLIA are not adequately tailored to the complexities of molecular diagnostics, and private-sector accreditation is voluntary and limited in scope. For most "high-complexity" tests, CMS has created specialty areas under CLIA, mandating, among other requirements, participation in specified proficiency testing programs. Molecular diagnostics are high-complexity tests, but CMS has not created a specialty area for these tests. In 2000, the Secretary's Advisory Committee on Genetic Testing concluded that oversight of genetics tests was insufficient to ensure their safety, accuracy, and clinical validity and recommended that CMS should develop a specialty area for genetic testing under CLIA. That recommendation has not been implemented and CMS recently asserted that there is insufficient "criticality" to warrant rule making for genetic testing.

METHODS AND PROCESSES NEEDED FOR CLINICAL EVALUATION AND ADOPTION

Diagnostic tests historically have been adopted into clinical practice with relatively little assessment of their value to patients and clinicians. That is slowly changing, as health care payors are beginning to demand more evidence of effectiveness when making coverage and reimbursement decisions. Appropriate clinical use of diagnostic tests requires assessments of the clinical risks and benefits, but the studies needed to make those assessments can be costly, lengthy, and difficult, making it hard for sponsoring companies to undertake them. New approaches to gathering data on effectiveness, cost, and value are needed to strike the necessary balance between encouraging innovation and ensuring that patients and providers have accurate and reliable tests. The final three recommendations suggest strategies to facilitate collection of needed data while also fostering expedient access to the market, and appropriate pricing of tests.

Recommendation 10:
CMS should modernize the process for evaluating, coding, and pricing diagnostic tests, and use the power of their longitudinal data to assess the value of tests.

- As previously recommended by an Institute of Medicine report (2000), Medicare ought to have "a single national, rational fee schedule" for clinical laboratory tests.
- Reimbursement policies should be clarified and the decision process should be made more uniform and transparent.
- CMS should convene stakeholders to develop consensus guidance on how to assess diagnostics to make coverage and reimbursement decisions.
- Expert panels should review new tests and reach consensus on coverage and pricing.

Rationale

Pricing of diagnostics is very different from that of drugs. For diagnostics, CMS uses "gap filling" and "cross-walking" to establish prices based on comparisons with tests and procedures already in use. Federal legislation also specifies national limitation amounts and links price increases to the consumer price index, whose rate of growth is below the rate of medical inflation. As a result, many experts argue that the reimbursement levels for diagnostics set by CMS do not adequately reflect their actual cost or clinical value, with some reimbursement rates too high relative to value, while others are too low.

Current pricing methods are particularly problematic for insurers with regard to homebrew tests. It is easier for payors to evaluate and control the use and the reimbursement rate of an FDA-approved test kit, which has its own individual Current Procedural Terminology (CPT) code. But for homebrew tests, a laboratory breaks down what they do into specific methods and analytes used, each with its own CPT code. Thus, a single test could entail 10–15 different codes, making it difficult to identify and evaluate appropriate use of the test.

Fair and rational pricing would foster innovation by enabling developers to better predict the return on investment. Seeking input from outside experts, as FDA does in evaluating new drug and device applications, would greatly improve the process. Although CMS is prohibited from using clinical value as a criterion for reimbursement, assessing the clinical value of tests would aid clinical decision making.

> **Recommendation 11:**
> **CMS (and other health care payors, including private insurers) should develop criteria for temporary, conditional coverage of new biomarker tests in certain circumstances to facilitate controlled and limited use of a diagnostic with a therapeutic and, even more importantly, a screening biomarker test, until sufficient evidence can be gathered to make an informed decision about standard (permanent, nonprovisional) coverage. Using this risk-sharing approach, payors would agree to provisionally cover new tests in specified circumstances with the proviso that, in the interim, data would be collected in conjunction with use of the test, to assess its clinical utility and value.**

Rationale

Biomarker tests often enter the market with little assessment or evidence of patient benefit, so payors may have to make coverage decisions with very little information. But premature adoption of inaccurate or ineffective biomarkers could be more costly to society than paying for conditional coverage, because, once provided, coverage is rarely retracted. Several national health care plans in Europe have high standards for evidence in coverage decisions, but they also commonly share some of the costs and risks of evidence development with technology sponsors. Coverage under Evidence Development (CED) is already being used by CMS in some cases (e.g., positron emission tomography imaging for cancer diagnosis, staging, and monitoring, as well as off-label uses of four drugs approved for colorectal cancer), demonstrating the feasibility of this approach. Thus, conditional coverage by CMS could provide a means of collecting important data on the use, effectiveness, and value of biomarker tests before they are broadly adopted. Private insurers may experience some difficulty in implementing conditional coverage because they are required to administer benefits according to the terms of their benefit plans, which often exclude experimental or investigational items, but it would be beneficial to examine and overcome these challenges. Because Medicare primarily covers patients who are older than 65, private insurers could make a very important contribution by collecting data on younger patient populations for whom cancer screening tests may yield the greatest gains in survival and reduced morbidity.

The conditional coverage approach would be especially useful for testing biomarkers for screening due to the very large populations needed to evaluate them. Companies do not have the financial means or incentives to undertake the very large, lengthy, and costly studies to assess a screening test. Most tests enter the market as a diagnostic but may then be adopted for screening, in the absence of adequate evaluation, via off-label use. Once adopted in such a fashion, it may be difficult or impossible to adequately assess the risks, benefits, and value of a screening test.

Recommendation 12:
When conditional coverage is applied, the cost-effectiveness of biomarkers should be studied by independent research entities, in conjunction with the assessment of technology accuracy and clinical effectiveness. This issue is particularly important for screening biomarkers due to the costs and potential morbidity of false positive tests.

- Optimal study designs are needed for high-quality population-based assessments of efficacy and cost-effectiveness of biomarker tests.
- To maximize the cost-effectiveness of diagnostics and therapeutics, it will be necessary to account for heterogeneity in patient populations, including risk, benefit, and behavior (e.g., patterns of use and self-selection).
- A structure and transparent process is needed for sharing information among laboratory test manufacturers, clinical laboratories, and health care payors. The format of the Academy of Managed Care Pharmacy evidence-based evaluation of drugs could be instructive in this regard.

Rationale

CMS is prohibited from using cost-effectiveness in coverage decisions, and currently, evidence for the application and utility of laboratory tests is often quite limited. However, demand for such evidence is increasing, so new methods and approaches are needed. The rapidly increasing costs of medical care are a common concern and are often attributed in part to adoption of new technologies. Although diagnostics account for only 1.6 percent of total Medicare costs, they influence the majority of downstream treatment decisions, so there is a need to assess potential indirect downstream risk and costs. In oncology, the cost of new targeted therapies can be an order of magnitude higher than traditional cancer treatments, and often only a fraction of patients benefit significantly from the treatments. Screening an entire asymptomatic population is also costly and can lead to harm as well as benefit. Biomarkers will be clinically valuable if they encourage appropriate selective use of treatments or identify cancers at a stage that is easier and less costly to treat.

1

Introduction

A long-standing goal of cancer research has been to identify the molecular mechanisms by which cancers develop, and then to detect those molecular markers of cancers early and to target those mechanisms with drugs specifically designed to attack them. A few remarkable strides have been made toward achieving that goal of "personalized medicine" in some cancers (The Royal Society, 2005). For example, the discovery of a chromosomal translocation in chronic myelogenous leukemia led to the development of a drug (imatinib) that targets the enzyme produced as a result of that translocation (reviewed by Druker, 2004; Baselga, 2006). In breast cancer, expression of the estrogen receptor serves as a biomarker for prognosis and identifies women who are likely to benefit from antiestrogen therapy (reviewed by Duffy, 2005; Ariazi et al., 2006). Similarly, the overexpression of HER2 (a growth factor receptor) in breast cancer serves as a biomarker for prognosis and for treatment with trastuzumab, a drug that targets that receptor's function (reviewed by Yeon and Pegram, 2005; Duffy, 2005; Baselga, 2006).

However, cancer is a collection of more than 100 different diseases, and, for most cancers, the molecular characteristics have not been fully classified and there are no known or validated markers for early detection, treatment planning, or targeted therapy. The diagnosis of cancers is still based largely on morphological examination of tumor biopsy specimens, as it has been for decades, but this approach has significant limitations for predicting a given tumor's potential for progression and response to treatment.

Recharacterization of cancers and other diseases in pathophysiological terms is key to the future of medicine. Considerable progress has been made in the molecular classification of some cancers, such as hematological malignancies (Box 1-1). More recently, the systematic analysis of genomic alterations in a small set of breast and prostate cancers revealed that individual tumors accumulate an average of approximately 90 mutant genes but that only a subset of these contribute to the neoplastic process. The authors identified 189 genes (average of 11 per tumor) that were mutated at

---

BOX 1-1   Biomarkers of Hematologic Cancers

The diagnosis of hematologic cancers presents an enormous challenge. The numerous stages of hematopoietic differentiation give rise to many biologically and clinically distinct cancers, most often via acquired genetic alterations. Knowledge of the biology underlying hematological malignancies has greatly increased in recent decades, leading to a much more sophisticated classification system that incorporates not only the traditional morphologic characteristics, but also immunophenotypic, genetic, and clinical features. However, even with this added information, considerable heterogeneity still exists within identified subtypes, with different clinical presentations and outcomes.

Researchers have long sought a classification system based on molecular pathogenesis, and DNA microarrays have been recently used to survey the expression of thousands of genes in parallel. Studies have identified novel disease subtypes and have also uncovered relationships between diseases previously considered to be unrelated. The results show great promise for refining diagnosis and prognosis, predicting response to treatment, and identifying potential targets for novel therapeutic interventions, although much work remains to be done before such tests can be routinely used to aid clinical decisions. For example, further clinical validation in larger cohorts and independent studies are needed, as well as test platform standardization and analytical validation. It is also not yet clear whether whole gene expression patterns are required, or whether a small set of genes will be sufficient to predict prognosis.

SOURCES: Reviewed by Staudt, 2003; Levene et al., 2003; Bullinger, 2005; Bullinger et al., 2005.

significant frequency, the vast majority of which were not previously known to be altered in tumors and are predicted to affect a wide range of cellular functions, including transcription, adhesion, and invasion, thus providing potential new targets for diagnosis and therapy, as well as new directions for basic research in tumor biology (Sjoblom et al., 2006). However, much work remains to be done.

Developing drugs and determining appropriate therapy for most diseases is still largely empirical and lacking well-defined molecular targets, and most medicines have been shown to be effective in less than 60 percent of patients in the disease populations that they address (Spear et al., 2001; Austin and Babbiss, 2006). In oncology, that figure is much lower, with an average drug response rate of less than 25 percent, due to the tremendous heterogeneity among patients with a given type of cancer. In addition, most cancer drugs are toxic agents that affect cell growth, so they often have significant side effects due to their activity against normal tissues in the body. In oncology, then, there is considerable opportunity for improving the drug development process as well as improving prevention, early detection, diagnosis, and treatment of cancers.

In principle, biomarkers should improve patient outcomes by ensuring that each patient receives the drugs that are most likely to be effective for his or her particular tumor, thereby enhancing the drug response rate and limiting toxicity. In addition to improving the effectiveness of therapy, biomarkers have the potential to improve the cost-effectiveness of treatment, both by avoiding the use of costly therapies to which a cancer will not respond and by avoiding the need to manage associated side effects of such treatments. Biomarkers that detect cancers at their earliest and most treatable stages should also improve patient outcome and the cost-effectiveness of therapy.

Yet despite years of research, the number of cancer biomarkers in clinical use is quite small (Duffy, 2005; Hayes, 2005; Gasparini et al., 2006). Although in recent decades, knowledge about the biology of cancers has increased greatly, and many candidate biomarkers have been reported, very few have been sufficiently validated to justify their use in developing drugs or making patient care decisions. Why is this so? The discovery and development of useful biomarkers pose enormous challenges, and many different factors contribute to the slow pace of biomarker development (FDA, 2004).

A discussion of how to develop and use biomarkers should start with a definition of the term. In its broadest sense, a biomarker is any biological, chemical, or biophysical indicator of an underlying biological process. From a medical perspective, a biomarker is a physiological characteristic that is indicative of health and disease; it has been explicitly defined as "a characteristic that is objectively measured and evaluated as an indicator of normal biologic processes, pathogenic processes, or pharmacologic response(s) to a therapeutic intervention" (Biomarkers Definitions Working Group, 2001). A cancer biomarker has been defined as "a molecular, cellular, tissue, or process-based alteration that provides indication of current, or more importantly, future behavior of cancer" (Hayes et al., 1996). These biological and physiological indicators could include a broad range of biochemical entities, such as nucleic acids, proteins, sugars, lipids, and small metabolites, as well as whole cells, in either specific tissues of interest or in the circulation. Detection of biomarkers, either individually or as larger sets or patterns, can be accomplished by a wide variety of methods, ranging from biochemical analysis of blood or tissue samples to biomedical imaging.

In fact, there are strong interconnections between biomedical imaging and the development of biomarkers. For example, biomarkers are increasingly needed to expand the capabilities of imaging, but imaging is also an important tool for validating biomarkers for specific uses. In addition, imaging will be necessary to identify the location and extent of tumors whose presence might be indicated by future biomarker tests. Although imaging is likely to play an increasingly important role in the future of cancer detection and therapy, the primary focus of this report is in vitro diagnostics.[1] Many of the challenges in biomarker development are relevant to both biomedical imaging and in vitro diagnostics, but in vivo imaging also entails a set of unique considerations (reviewed by Chandra et al., 2005), in part because it often requires injection of chemical agents, and thus it faces some of the same challenges as drug development but lacks the financial incentives of the drug industry. These issues are not addressed in the report, but this topic will be covered in more detail in a future workshop on drug development that is tentatively being planned by the IOM's National Cancer Policy Forum.

Biomarkers can be useful at any point in the biomedical continuum, from basic biomedical research through pharmaceutical discovery and preclinical development, clinical trials, and patient care (Kiviat and Critchlow,

---

[1] In this report, "diagnostic" is often used synonymously with "biomarker test." These terms refer to any laboratory-based test that can be used in drug discovery and development as well as in patient care and clinical decision making.

2002; Srivastava and Wagner, 2002; Nakamura and Grody, 2004; Park et al., 2004; Floyd and McShane, 2004; Caprioli, 2005; Ludwig and Weinstein, 2005; Dalton and Friend, 2006, Kelloff et al., 2006; Weissleider, 2006). Clinical applications include disease risk stratification, chemoprevention, disease screening, diagnosis and prognosis/prediction, treatment planning and monitoring, and posttreatment surveillance (Table 1-1). For drug development, biomarkers may be used to assess drug candidates for evidence of safety and efficacy at each step of the development process (Table 1-2).

Two primary challenges to developing cancer biomarkers are the discovery of candidate markers and the validation of those candidates for specific uses. The discovery process depends on the technologies available to interrogate the complex biochemistry of health and disease in order to identify differences that can be detected consistently in diverse populations.

TABLE 1-1 Use of Cancer Biomarkers in Patient Care

Clinical Biomarker Use	Clinical Objective
Risk stratification	Assess the likelihood that cancers will develop (or recur)
Chemoprevention	Identify and target molecular mechanisms of carcinogenesis in precancerous tissues
Screening	Detect and treat early-stage cancers in the asymptomatic population
Diagnosis	Definitively establish the presence of cancer
Classification	Classify patients by disease subset
Prognosis	Predict the probable outcome of cancer regardless of therapy, to determine the aggressiveness of treatment
Prediction/ treatment stratification	Predict response to particular therapies and choose the drug that is mostly likely to yield a favorable response in a given patient
Risk management	Identify patients with a high probability of adverse effects of a treatment
Therapy monitoring	Determine whether a therapy is having the intended effect on a disease and whether adverse effects arise
Posttreatment surveillance	Early detection and treatment of recurrent disease

TABLE 1-2 Use of Biomarkers in Drug Development

Biomarker Use	Drug Development Objective
Target validation	Demonstrate that a potential drug target plays a key role in the disease process
Early compound screening	Identify compounds with the most promise for efficacy and safety
Pharmacodynamic assays	Determine drug activity; select dose and schedule
Patient selection	In clinical trials, patient selection (inclusion/exclusion) by disease subset or probability of response/adverse events
Surrogate endpoint	Use of a short-term outcome measure in place of the long-term primary endpoint to determine more quickly whether the treatment is efficacious and safe in drug regulatory approval

Recent technological developments, especially in genomics and proteomics, have made it much easier to examine a large number of potential markers at once. Nonetheless, progress is still limited by the sensitivity and specificity of the current technologies, as well as the methods and tools used to analyze the enormous pools of data generated by high-throughput technologies, and there is still a need for new and improved technologies to discover potential biomarkers.

The validation process is also arduous and costly, often requiring collection of or access to many patient samples with extensive clinical annotation and long-term follow-up. In addition, a biomarker must be validated for each specific application (as in Tables 1-1 and 1-2) for which it will be used. For example, the criteria for validating a biomarker for use as a screening test in asymptomatic populations will be quite different from those used to validate a biomarker for use as a surrogate end point in clinical trials of a drug, since the applications are so fundamentally different. There must be convincing evidence that a surrogate end point accurately predicts the clinical endpoint of interest. In the case of screening, a test must have sufficient sensitivity, specificity, and positive predictive value[2] to accurately identify a disease in the general population.

---

[2]The probability that an individual with a positive test has a particular disease, or characteristic, that the test is designed to detect. It is a measure of the ratio of true positives to (false + true positives).

Furthermore, health care payors have begun to require more data on the clinical validity of tests when making decisions about coverage and reimbursement. However, there is a lack of standards and guidelines for how to assess biomarkers and determine appropriate usage. These topics and the associated challenges are covered in more detail in Chapters 2 and 3.

Clearly, much remains to be done to achieve the vital goal of effective early detection and individualized therapy for all people with cancer. A major research investment will be required to accomplish that goal. The opportunity cost of further progress in the field is always a consideration, as there are many competing needs and goals in biomedical research. However, a considerable investment is already being made in this field of research, and much could be accomplished by improving the discovery and development process to make the most of both ongoing and future efforts. The recommendations put forth by the committee in this report strive to realize these advances.

COMMITTEE CHARGE

The Committee on Developing Biomarker-based Tools for Cancer Screening, Diagnosis, and Treatment was asked to address (1) the potential to improve cancer screening, diagnosis, and therapy through the use of emerging biomarker technologies; (2) current limitations of genomics and proteomics technologies for cancer detection, diagnosis, and drug development, as well as steps that could be taken to overcome these limitations; (3) the logistics and cost of coordinating the development of biomarkers and targeted therapies; (4) regulatory oversight of biomarker development and use; (5) the adoption of biomarker-based tests and therapeutics into clinical practice; and (6) some of the potential economic implications of adopting these emerging technologies.

A workshop hosted by the National Cancer Policy Forum in March 2006 addressed a similar set of questions, and the proceedings of the meeting (IOM, 2006, see Appendix) served as a primary input to the committee's deliberations.

FRAMEWORK OF THE REPORT

Following this introduction, **Chapter 2** provides a brief overview of the technologies and methods used to discover and develop biomarkers for pre-clinical and clinical use. It describes several resources for and approaches to biomarker discovery and development that the committee agreed warranted further attention, including biorepositories, consortia, and demonstration projects.

Chapter 3 reviews current oversight of biomarker development and use by federal agencies (Food and Drug Administration and Centers for Medicare and Medicaid Services). It examines a variety of approaches to improve the development and evaluation process and to ensure the quality of biomarker tests used by patients and physicians while also fostering innovation.

Chapter 4 provides a brief overview of the technology evaluation and adoption processes and examines possible ways to facilitate data collection and analysis to monitor and improve the value of biomarker tests.

REFERENCES

Ariazi EA, Ariazi JL, Cordera F, Jordan VC. 2006. Estrogen receptors as therapeutic targets in breast cancer. *Current Topics in Medicinal Chemistry* 6(3):181-202.

Austin M, Babbiss L. 2006. Commentary: When and how biomarkers could be used in 2016. *AAPS Journal* 8(1):E185-E189.

Baselga J. 2006. Targeting tyrosine kinases in cancer: The second wave. *Science* 312(5777): 1175-1178.

Biomarkers Definitions Working Group. 2001. Biomarkers and surrogate endpoints: Preferred definitions and conceptual framework. *Clinical Pharmacology and Therapeutics* 69:89-95.

Bullinger L. 2005. Gene expression profiling in acute myeloid leukemia. *Haematologica* 91(6):733-738.

Bullinger L, Doner H, Pollack JR. 2005. Genomics in myeloid leukemias: An array of possibilities. *Reviews in Clinical and Experimental Hematology* 9(1):E2.

Caprioli RM. 2005. Deciphering protein molecular signatures in cancer tissues to aid in diagnosis, prognosis, and therapy. *Cancer Research* 65(23):10642-10645.

Chandra S, Muir C, Silva M, Carr S. 2005. Imaging biomarkers in drug development: an overview of opportunities and open issues. *Journal of Proteome Research* 4(4):1134-1137.

Dalton WS, Friend SH. 2006. Cancer biomarkers—an invitation to the table. *Science* 312(5777): 1165-1168.

Druker BJ. 2004. Imatinib as a paradigm of targeted therapies. *Advances in Cancer Research* 91:1-30.

Duffy MJ. 2005. Predictive markers in breast and other cancers: A review. *Clinical Chemistry* 51(3):494-503.

FDA (Food and Drug Administration). 2004. *Challenge and Opportunity on the Critical Path to New Medical Products.* [Online]. Available: http://www.fda.gov/oc/initiatives/criticalpath/whitepaper.html [accessed September 2006].

Floyd E, McShane TM. 2004. Development and use of biomarkers in oncology drug development. *Toxicologic Pathology* 32 Suppl 1:106-115.

Gasparini G, Longo R, Torino F, Gattuso D, Morabito A, Toffoli G. 2006. Is tailored therapy feasible in oncology? *Critical Reviews in Oncology/Hematology* 57(1):79-101.

Hayes DF. 2005. Prognostic and predictive factors revisited. *Breast* 14(6):493-499.

Hayes DF, Bast RC, Desch CE, Fritsche H Jr, Kemeny NE, Jessup JM, Locker GY, Macdonald JS, Mennel RG, Norton L, Ravdin P, Taube S, Winn RJ. 1996. Tumor marker utility grading system: A framework to evaluate clinical utility of tumor markers. *Journal of the National Cancer Institute* 88(20):1456-1466.

IOM (Institute of Medicine). 2006. *Developing Biomarker-Based Tools for Cancer Screening, Diagnosis, and Treatment: The State of the Science, Evaluation, Implementation, and Economics. A Workshop Summary.* Patlak M, Nass S, rapporteurs. Washington, DC: The National Academies Press.

Kelloff GJ, Lippman SM, Dannenberg AJ, Sigman CC, Pearce HL, Reid BJ, Szabo E, Jordan VC, Spitz MR, Mills GB, Papadimitrakopoulou VA, Lotan R, Aggarwal BB, Bresalier RS, Kim J, Arun B, Lu KH, Thomas ME, Rhodes HE, Brewer MA, Follen M, Shin DM, Parnes HL, Siegfried JM, Evans AA, Blot WJ, Chow WH, Blount PL, Maley CC, Wang KK, Lam S, Lee JJ, Dubinett SM, Engstrom PF, Meyskens FL Jr, O'Shaughnessy J, Hawk ET, Levin B, Nelson WG, Hong WK. 2006. Progress in chemoprevention drug development: The promise of molecular biomarkers for prevention of intraepithelial neoplasia and cancer—A plan to move forward. *Clinical Cancer Research* 12(12):3661-3697.

Kiviat NB, Critchlow CW. 2002. Novel approaches to identification of biomarkers for detection of early stage cancer. *Disease Markers* 18:73-81.

Levene AP, Morgan GJ, Davies FE. 2003. The use of genetic microarray analysis to classify and predict prognosis in haematological malignancies. *Clinical and Laboratory Haemotology* 25:209-220.

Ludwig JA, Weinstein JN. 2005. Biomarkers in cancer staging, prognosis and treatment selection. *Nature Reviews* 5:845-856.

Nakamura R, Grody W. 2004. Cancer diagnostics: Current and future trends. In: Nakamura RM, Grody WW, Wu JT, Nagle RB, *General Considerations in the Use and Application of Laboratory Tests for the Evaluation of Cancer.* Totawa, NJ: Humana Press Inc. Pp. 3-14.

Park JW, Kerbel RS, Kelloff GJ, Barrett JC, Chabner BA, Parkinson DR, Peck J, Ruddon RW, Sigman CC, Slamon DJ. 2004. Rationale for biomarkers and surrogate end points in mechanism-driven oncology drug development. *Clinical Cancer Research* 10(11):3885-3896.

The Royal Society. 2005. Personalised Medicines: Hopes and Realities. London, UK: The Royal Society.

Sjoblom T, Jones S, Wood LD, Parsons DW, Lin J, Barber TD, Mandelker D, Leary RJ, Ptak J, Silliman N, Szabo S, Buckhaults P, Farrell C, Meeh P, Markowitz SD, Willis J, Dawson D, Willson JK, Gazdar AF, Hartigan J, Wu L, Liu C, Parmigiani G, Park BH, Bachman KE, Papadopoulos N, Vogelstein B, Kinzler KW, Velculescu VE. 2006. The consensus coding sequences of human breast and colorectal cancers. *Science* 314:268-274.

Spear BB, Heath-Chiozzi M, Huff J. 2001. Clinical application of pharmacogenetics. *Trends in Molecular Medicine* 7:201-204.

Srivastava S, Wagner J. 2002. Surrogate endpoints in medicine. *Disease Markers* 18:39-40.

Stuadt LM. 2003. Molecular diagnosis of the haematologic cancers. *New England Journal of Medicine* 348:1777-1785.

Weissleder R. 2006. Molecular imaging in cancer. *Science* 312:1168-1171.

Yeon CH, Pegram MD. 2005. Anti-erbB-2 antibody trastuzumab in the treatment of HER2-amplified breast cancer. *Investigational New Drugs* 23(5):391-409.

2

Methods, Tools, and Resources Needed to Discover and Develop Biomarkers

OVERVIEW OF THE BIOMARKER DISCOVERY AND DEVELOPMENT PROCESS

Scientists have been searching for cancer biomarkers for many years, but the methods of discovery have changed as new technologies have been developed. Traditionally, scientists have relied on conventional laboratory research tools, such as gel electrophoresis and immunohistochemistry, to identify altered genes and changes in mRNA and protein expression (Ross et al., 2004). Progress in this work has been slow because researchers could examine only one or a small number of candidate markers at a time, and the methods required some prior knowledge and experience with the potential markers of interest. More recently, many novel high-throughput technologies (Kiviat and Critchlow, 2002; Fan et al., 2004; Aebersold et al., 2005; Weckwerth and Morgenthal, 2005; De Bortoli and Biglia, 2006; IOM, 2006a), especially in the fields of genomics and proteomics, have made it easier to interrogate hundreds or even thousands of potential biomarkers at once, without prior knowledge of the underlying biology or pathophysiology of the system being studied (Table 2-1). As a result, there has been a flood of new data and renewed interest in discovering novel biomarkers for use in drug development and patient care.

The goal of these discovery methods is to identify genetic variations or mutations as well as changes in gene or protein expression or activity that can be linked to a disease state or a response to a medical intervention.

TABLE 2-1 Examples of Biomarker Categories and High-Throughput Methods of Discovery

Biomarker Category	Examples of Methods
Genomics	
DNA-based	
Copy number/loss of heterozygosity	Various DNA arrays
Sequence variation	Various sequencing methods
Epigenetic variation	
Genome rearrangements	
RNA-based	
mRNA signatures	Various DNA arrays
miRNA signatures	
Proteomics	Mass spectrometry
Proteins	Liquid chromatogrpahy
Peptides	Protein arrays
Metabolomics	
Metabolites	Mass spectrometry
Lipids	Liquid chromatography
Carbohydrates	Nuclear magnetic resonance

SOURCE: Derived from IOM, 2006a.

Analysis of these large datasets requires sophisticated algorithms and bioinformatics to identify individual markers of interest or to derive signatures or patterns of many markers (reviewed by Cristoni and Bernardi, 2004; Englbrecht and Facius, 2005; Tinker et al., 2006). Although these methods are continually evolving and being improved, there is still a great need for novel approaches to data analysis, especially with regard to network oriented models that can incorporate many different types of data to fully integrate the vast complexity of biology in health and disease. However, identifying biomarker patterns or specific changes in genes or the products of gene expression in tumors is only the beginning of the process to develop cancer biomarkers.

Before a candidate biomarker can be put into use, it must undergo several stages of confirmation, validation, and qualification for use (Wagner, 2002; Feng et al., 2004; Ransohoff, 2004, 2005; Simon, 2005; De Bortoli and Biglia, 2006). Analytical validation is the process of assessing the assay or measurement performance characteristics, while qualification is the evidentiary process of linking a biomarker with the biology and clinical endpoints (that is, clinical validity and utility). Both are intended to ensure that a biomarker is fit for a specified purpose, but it is often not clear how best to prove the performance characteristics of a biomarker-based test, especially for those that employ newer technologies, since many lack a gold standard for comparison (IOM, 2006a). Ultimately, a test used to make clinical decisions must, in combination with an intervention, lead to a beneficial impact on patient outcomes. Thus, use as a clinical diagnostic also involves evaluations of benefit, harm, cost, and effort (Ransohoff, 2004).

Once an appropriate method has been selected for measuring the biomarker or pattern of markers, the technical parameters of the test must be well defined to establish sensitivity, specificity, reproducibility, and reliability of the measurements. However, different technology platforms may be needed at different stages of biomarker development. Platforms for biomarker discovery generally need to process many biomarkers simultaneously, but they can be low throughput in terms of the samples processed. Platforms for clinical research need to be high throughput in terms of specimen processing, but they usually focus on a smaller number of markers. Platforms for clinical practice need to be inexpensive and robust, and ideally results should be easily and objectively quantifiable.

The validity of the biomarker as an indicator of a biological, pathological, or pharmacological process must also be confirmed in carefully designed studies. Validation is necessary for each potential use of a biomarker, and the level of evidence needed to implement a biomarker varies with the intended use. For example, in the drug development process, biomarkers can play a role at many different stages, from early, exploratory research to surrogacy for a clinical endpoint in large clinical trials, and the required degree of validation increases along that continuum (Table 2-2; see also Table 1-2). In the drug development process, the highest level of evidence is required if the biomarker is to be used as a surrogate endpoint—it must be qualified for that specific use by clearly demonstrating in clinical studies that the marker accurately predicts the clinical endpoint of interest. Correlation and plausibility is not sufficient (Srivastava and Wagner, 2002; Wagner, 2002; Fleming, 2005). For instance, tumor shrinkage might seem to be a plausible surrogate for treatment efficacy, but in fact, tumor shrinkage in response to a drug does not necessarily lead to improved patient survival (Norton, 1997; Citron, 2004; Hudis, 2005). Similarly, inhibition of pre-invasive abnormalities is widely thought to predict a reduction in invasive cancer (Kelloff et al., 2006), but the relationship has not been fully validated in clinical studies, and a recent study even showed that the antiestrogen

TABLE 2-2 Biomarker Validation and Qualification Requires
Demonstration of Fitness for a Specified Purpose

Type of Biomarker	Definition	Purpose
Exploration	Research and development tool	Hypothesis generation
Demonstration	Probable or emerging biomarker	Decision making, supporting evidence with primary clinical evidence
Characterization	Known or established biomarker	Decision making, dose finding, secondary/ tertiary claims
Surrogacy	Biomarker can substitute for a clinical endpoint	Regulatory approval

NOTE: Shown are four categories of biomarkers used for drug development and their intended purpose.
SOURCE: Adapted from Wagner, 2006.

raloxifene can reduce the risk of invasive breast cancer without significantly reducing the incidence of ductal carcinoma in situ, a preinvasive lesion with potential to develop into invasive cancer (NSABP, 2006).

Similarly, the criteria for validation vary among the many possible clinical uses of biomarkers (see Table 1-1). For example, validation of a biomarker for screening, which entails the systematic testing of an asymptomatic population to identify evidence of particular type of cancer, requires proof that the biomarker detects the disease with a high degree of sensitivity, specificity, and positive predictive value. Ultimately, the clinical value of a screening test will also also depend on whether the routine use of the test, combined with appropriate interventions, reduces the morbidity and mortality due to that disease. In contrast, the clinical validation criteria for a diagnostic biomarker, which is used to definitively determine the presence or absence of cancer in patients with symptoms or a known abnormality, are less onerous.

The vast majority of candidate biomarkers never progress past the initial discovery phase, and very few become qualified as surrogate endpoints or useful clinical tests, in part because further evaluation is expensive and time-consuming, with uncertain outcome (and thus a risky endeavor)

(Hayes et al., 1998; Wagner, 2002; Srivastava and Wagner, 2002; Feng et al., 2004; Altman and Riley, 2005; Fleming, 2005; Simon, 2005; IOM, 2006a, 2006b). However, it has also been argued that the rules of evidence to assess the validity of biomarkers are both underdeveloped and not routinely applied (Ransohoff, 2004; Altman and Riley, 2005; LaBaer, 2005). The mechanisms of disease and pharmacologic response are complex and challenging to discern, and lack of appropriate study designs and analytic methods exacerbates the challenge. Perhaps as a result, both genomic and proteomic studies of the same cancer types have often identified discordant biomarker candidates and patterns (reviewed by Diamandis, 2004; Dalton and Friend, 2006; Quackenbush, 2006).

Because of the enormous number of genes and proteins analyzed in genomic and proteomic studies, many false findings can be expected unless appropriate statistical methods are used (Simon, 2005). For example, in the discovery setting, overfitting can lead to erroneous identification of markers or patterns in association with disease. This occurs when multivariate analysis is used to assess associations between large numbers of possible predictors and an outcome—that is, a pattern is found that fits perfectly, but by chance (Ransohoff, 2004, 2005; Simon, 2005, 2006). Overfitting can be easily identified by checking reproducibility in a separate, independent group of individuals, but most published studies do not report this essential assessment of reproducibility (Ransohoff, 2004, 2005).

Sample bias can also render conclusions drawn from a biomarker study invalid (Ransohoff, 2005). Bias can be defined broadly as the systematic but unintentional erroneous association of some characteristic with a group in a way that distorts a comparison with another group. The design, conduct, and interpretation of randomized clinical trials to assess medical interventions place high importance on ensuring that the treated and untreated patient populations are similar in every respect except for the treatment to avoid biases that could affect the outcome and thus the conclusions drawn from the results. However, most research on molecular markers for diagnosis or prognosis entails observational studies, in which it is difficult or impossible to ensure or even fully assess the similarity of the comparison groups, and which are much more likely to result in biased conclusions as a result. In fact, Ransohoff has suggested that bias can be so powerful in non-experimental observational research that a study should be presumed biased until proven otherwise (2005). He notes that a single bias might produce errors sufficiently large to invalidate results. Thus, great care must be taken in the design, conduct, interpretation, and reporting of such research.

The validity of biomarker research also depends on the generalizability of the results. Generalizability concerns how broadly the results can be applied and depends on the characteristics of study participants and how they were selected (Ransohoff, 2005). Initial studies often establish proof of principle, but they have limited generalizability. Subsequent larger studies then aim to assess broader generalizability.

However, the developmental sequence for biomarker development is much less well defined than for drug development. In drug development, the phases of research are clearly delineated in a step-wise fashion to examine several key issues, including generalizability. Phase I studies aim to establish appropriate doses and to identify potential side effects, while phase II studies begin to address biological activity and adverse events. Phase III studies are larger and seek more definitive conclusions on efficacy and safety. Patients in phase I and II studies have often failed all available treatment options and have diverse characteristics, whereas phase III trials select patients who are more representative of how the drug will actually be used in clinical practice (Ransohoff, 2005).

Delineating the phases for biomarker development is more difficult, in part because biomarker tests can be used for many different purposes, and thus research to assess the usefulness of a test must be designed to examine specific applications. The variability in technologies used to identify biomarkers further complicates the situation. As a result, proposals to establish developmental phases for biomarkers have focused on specific uses, such as early detection or surrogate endpoints, or specific methods (Pepe et al., 2001; Baker et al., 2004; Zolg and Langen, 2004). Perhaps because of this variability in the development pathway, the level of assessment and oversight for biomarkers is also more variable, and usually less stringent, than for drugs. The process of developing and implementing biomarkers differs from that of drugs in other ways as well, including economically. These issues are covered in more detail in Chapters 3 and 4.

THE NEED FOR NEW, INNOVATIVE TECHNOLOGIES

If the full potential of cancer biomarker-based tools in early detection, treatment, and drug development is to be realized, it will be important to optimize efforts to discover and validate putative biomarkers. Progress in biomarker discovery and development is directly dependent on the capacities of the technologies available. Initially, high-throughput methods to discover biomarkers and expression patterns focused on nucleic acids, because the methods were more advanced and fully characterized than for other cellular components. The Human Genome Project[1] spurred interest in the development and application of methods to access nucleic acids, and although there is still a need for standardization of reagents, platforms, and analyses, considerable progress has been made in the field. In addition, ongoing work to sequence the genomes of individual tumors as well as other organisms continues to spur the development of new technologies. For example, single-molecule sequencing[2] is likely to lower the cost of sequencing significantly, and should reduce the problems that arise from normal cell contamination of tumor samples (IOM, 2006a).

However, there are limitations to what can be learned from genomics approaches to assess DNA and RNA. Although RNA assays can detect dynamic changes in gene expression as well as identify upstream DNA-level abnormalities in cancers, it is more costly and difficult to work with than the comparatively stable DNA. It can be argued that proteins, which perform many essential functions in cells, may provide more meaningful biomarkers than either type of nucleic acid because changes in DNA and RNA are not always directly linked to altered protein expression, modification, or function. But progress in the identification of protein biomarkers has lagged, in part because proteins are more numerous and far more subject to quantitative and post-translational structural changes than genes, and in part because of the limitations of current technologies (Tyers and Mann, 2003; Aebersold et al., 2005; Hartwell, 2005; Cottingham, 2006;). Technologies used to examine other types of biomarkers, such as metabolomics, are even less developed and characterized. Metabolomics entails the study of metabolic responses to drugs, environmental changes, and diseases via identification of small-molecule metabolite profiles; that is, it attempts to measure the metabolic consequences of altered genes and protein expression.

---

[1]The Human Genome Project was an international research project to map each human gene and to completely sequence human DNA. Approximately $2.7 billion were invested in the project between 1990 and 2003.

[2]Single-molecule sequencing, also called nanopore sequencing, is a method for sequencing DNA that involves passing the DNA through small pores about 1 nanometer in diameter. The size of the pore ensures that the DNA is forced through the hole as a long string, one base at a time. The base (i.e., adenine, guanine, cytosine, or thymine) is identified by the characteristic obstruction it creates in the pore, which is detected electrically. Single-molecule sequencing can be a more sensitive technique for identifying relatively rare genetic strands in a sample, without the need for replicating them with a polymerase chain reaction.

Proteomics research aims to interrogate extremely complex protein mixtures in blood and tissues. It has been estimated that blood contains more than 100,000 different protein forms with abundances that span 10–12 orders of magnitude (Anderson and Anderson, 2002; Jacobs et al., 2005). The leading high-throughput proteomics technology, mass spectrometry (MS), has limited ability to identify and quantify proteins in complex mixtures. Many known biomarkers occur at very low abundance and would not be identified by current technologies (Aebersold et al., 2005; Jacobs et al., 2005; Kolch et al., 2005). New fractionation methods (for depletion or enrichment) prior to MS could improve the process, as currently available methods are tedious and expensive, and are not amenable to high-throughput analysis. Furthermore, identification of the various peptides or proteins detected by the technology remains a difficult challenge. Antibodies raised against specific protein biomarkers could fill this gap, but currently antibodies are not available for most of the proteins that could be disease biomarkers (Anderson and Anderson, 2002; Aebersold et al., 2005; Hartwell, 2005; Jacobs et al., 2005; Cottingham, 2006).

In addition to improving the sensitivity, specificity, and dynamic range of these technologies, it will be important to process the resultant data efficiently and effectively, by developing new software packages, algorithms, and statistical and computation models, including those that can integrate data from multiple inputs, such as proteomic and genomic data from the same samples (Cristoni and Bernardi, 2004; Englbrecht and Facius, 2005; Tinker et al., 2006). These approaches will also necessitate novel sample preparation procedures, from a variety of sources such as blood, plasma, tissues, and cells, and for a variety of analytic technologies, including metabolomics, proteomics, and genomics. Finally it will be necessary to develop new assays to translate discovery into viable clinical tests, and to develop novel approaches for real-time in vivo detection, via imaging and nanomaterials. The Human Proteome Organisation (HUPO), an international consortium of national proteomics research associations, government researchers, academic institutions, and industry partners, has begun to examine some of these issues in a pilot phase of its Plasma Proteome Project (Omenn et al., 2005), and progress is being made on several fronts (de Hoog and Mann, 2004; Ong and Mann, 2005), but much work remains to be done. There is a significant need for new and improved technologies for biomarker discovery and development, particularly in the field of proteomics. Such new technologies also will yield dividends in improved capabilities for understanding fundamental cellular processes in cancers and systems biology in general.

Although technology development and directed discovery have not traditionally been a primary focus of National Institutes of Health (NIH) funding, and the NIH peer review process generally does not favor high-risk projects (IOM, 2003), that has been changing in recent years and there are some precedents for funding initiatives that focus on large-scale discovery projects and also catalyze the development of improved technologies for biomedical research. As noted above, the Human Genome Project drove not only the development of new technologies, but also improvements in the existing technologies through automation, data standards, and quality control (Aebersold et al., 2005; Hartwell, 2005). More recently, the NIH Roadmap proposed a framework for the priorities NIH as a whole must address in order to optimize its entire research portfolio, laying out a vision for a more efficient and productive system of medical research (NIH 2006b). The NIH Roadmap identified opportunities in three main areas: new pathways to discovery, research teams of the future, and re-engineering the clinical research enterprise. The first main area, pathways to discovery, aims to advance the quantitative understanding of complex biological systems by deciphering the many interconnected networks of molecules that comprise cells and tissues, their interactions, and their regulation. In addition the program aims to increase access in the research community to new and better technologies, databases and other scientific resources that are more sensitive, more robust, and more easily adaptable to evolving needs.

The Protein Structure Initiative (PSI), a $600 million, 10-year venture funded by the National Center for Research Resources of the National Institute of General Medical Sciences, is a recent example of an NIH program that explicitly funded technology development in the initial phase of the project. The overarching goal of the initiative is to determine the structure and function of thousands of proteins by 2010, with the final product serving as an inventory of all the protein structure families in nature. In the first phase of the initiative, PSI funded nine research centers that focused on developing novel and innovative approaches and technologies, such as robotic instruments, to determine protein 3-D structures from knowledge of their amino acid sequences (NIGMS, 2006). Technological innovations were developed for each step of the process, from the initial target selection, to the final poststructural analysis. According to NIH leadership, PSI succeeded in reducing costs four-fold from the initial year, and the technological improvements are likely to have a broad impact on protein structure research, beyond the PSI-funded work (Norvell and Berg, 2005).

NCI currently also sponsors development of novel nanotechnologies as tools to accelerate advances in biomarker research. Clinical applications of nanotechnologies include measurement and analysis of biomarkers *in vivo* for early cancer detection, prevention and monitoring of treatment response. NCI programs such as the Innovative Molecular Analysis Technologies Program provide funding for research projects to develop new and emerging technologies, including nanotechnology methods and tools, and for refining existing technologies through to development and commercialization. In September 2004, NCI announced a 5-year $144.3 million initiative to develop nanotechnologies to be used in cancer research (NCI, 2004). The goals of this initiative, the NCI Alliance for Nanotechnology in Cancer, include the development of research tools to identify new biological targets, agents to monitor predictive molecular changes and prevent precancerous cells from becoming malignant, imaging agents and diagnostics to detect cancers early, and systems to provide real-time assessments of therapeutic and surgical efficacy.[3] Emerging nanotechnologies include quantum dots, gold nanoparticles, and cantilevers. Quantum dots and magnetic nanoparticles can be used for barcoding of specific analytes, and gold and magnetic nanoparticles are components of a possible alternative to PCR known as the bio-barcode assay. Nanotechnologies can be used to genotype at high-throughput, and some researchers believe that they have the potential to reduce cost for many diagnostic applications (Azzazy et al., 2006). In addition, the size of nanoparticles makes them compatible with *in vivo* molecular manipulation and measurement (Yezhelyev et al., 2006). Nanodiagnostic assays have already been used to detect Alzheimer biomarkers in cerebrospinal fluid (Azzazy et al., 2006).

The National Cancer Institute (NCI) has also recently launched new funding initiatives for proteomics research, noting that "current proteomic technology approaches are insufficient to reliably and reproducibly discover, identify, and quantify peptides and proteins of clinical significance for cancer" from complex patient samples. The NCI's Clinical Proteomic Technologies Initiative for Cancer program is a 5-year $104 million program that includes two funding opportunities for proteomics technology development: Advanced Proteomic Platforms and Computational Sciences (DHHS, 2005) and Clinical Proteomic Technology Assessment for

---

[3]http://otir.cancer.gov/programs/ati_nano.asp.

Cancer (CPTAC) (DHHS, 2006b). The goal of the former is to improve technology for protein/peptide detection, recognition, measurement, and characterization in biological fluids and to develop computational, statistical, and mathematical approaches for the analysis, processing, and exchange of large proteomic datasets. The goal of the CPTAC is more specifically to improve measurements of proteins and peptides with mass spectrometry and affinity-based proteomics platforms. The CPTAC Request for Application notes that

> CPTAC teams will . . . be responsible for refining, comparing, and optimizing existing proteomics methods and applications. Improvements are sought in areas such as: sample collection and fractionation, and detection, identification, and quantification of proteins or peptides of interest. In addition, rigorous method/technology validations are needed to ensure reliable and reproducible results for proteomics analyses of complex biological mixtures. The CPTAC program is not designed for an explicit goal of developing new technologies and/or advanced applications. Nonetheless, since the priority in the initiative is the integration of the appropriate scientific expertise and infrastructure, the participating groups of scientists should be capable of efficiently implementing new technologies that might emerge during the life of the program.

While these NIH initiatives are to be commended, a review of projects funded through them suggests that many projects focus primarily on improving existing technologies, rather than on the development of completely novel technologies. To achieve the latter, it might be better to undertake a highly directed contract-based program. An examination of the Defense Advanced Research Projects Agency (DARPA) might prove instructive in that regard. DARPA is the central research and development organization for the Department of Defense, and it has focused on research projects that are high risk but also have potential for high payoff if successful (Box 2-1; IOM, 2003). As such, this approach is particularly amenable to technology development, and past leaders of NIH and NCI have expressed interest in adopting some aspects of the DARPA model to spark technological innovation. In fact, under the leadership of former NCI director Richard Klausner, NCI launched a pilot program that was modeled in part after DARPA, as well as other agencies, including the National Aerospace and Space Administration. Established in 1999, the Unconventional Innovations Program (UIP) focused on the development of novel, long-range technologies to support cancer research. The Program funded research through contracts instead of grants, allowing for enforcement of deadlines for specific milestones along the research track in order for researchers to

BOX 2-1   Overview of DARPA

The Defense Advanced Research Projects Agency (DARPA) agency was created in 1958 on the following founding principles, which are still adhered to:

- Small and flexible establishment.
- Flat organization.
- Substantial autonomy and freedom from bureaucratic impediments.
- Technical staff drawn from world-class scientists and engineers with representation from industry, universities, government laboratories, and federally funded research and development centers.
- Technical staff assigned for 3–5 years and rotated to ensure fresh thinking and perspectives.
- Project based: all efforts are typically 3–5 years long, with a strong focus on end-goals. Major technological challenges may be addressed over much longer times, but only as a series of focused steps. Projects are not renewed.
- Necessary supporting personnel (technical, contracting, administrative) are hired on a temporary basis to provide complete flexibility to undertake and abandon an area without problems of sustaining staff. Program managers (the heart of DARPA) are selected to be technically

continue to receive funds. UIP management actively recruited the interest and involvement of investigators from disciplines that have not traditionally received support from NCI and assembled interdisciplinary research teams focused on cancer detection technologies, including nanotechnologies (IOM, 2003; NCI, 1999).

Regardless of which funding model is used to foster innovative technology development, it will be essential to take an organized, comprehensive approach to the problem and to include a broad array of extramural experts in all aspects of program planning, execution, and oversight. For example, program managers with current expertise in the field could be recruited to direct the projects, similar to the DARPA approach. In addition, panels of experts should provide an oversight role in establishing goals and reviewing progress and performance of the program and individual contracts. Con-

outstanding and entrepreneurial. The best DARPA program managers have always been free-thinking zealots in pursuit of their goals.
- Management is focused on good stewardship of taxpayer funds but imposes little else in terms of rules. Management's job is to enable the program managers.
- A complete acceptance of failure if the payoff for success would have been high enough.

Best known for its role in developing the Internet, most of the work funded by DARPA has focused on computer and software development, engineering, materials science, microelectronics, and robotics, although more recently it has begun a limited program in basic molecular biology. With funding of approximately $150 M annually in recent years, DARPA's small group of expert program managers has extensive power to direct high-risk projects that would not normally fare well in peer review. The contracts with industry, academic, and government labs call for defined deliverables and allow less promising work to be canceled easily. The funded researchers often attend team meetings, file frequent reports, and work cooperatively with other contractors. DARPA has been particularly successful in forging new directions of research to create new fields and in solving specific technical problems by fostering the development of new technologies.

SOURCES: Adapted from IOM, 2003; DARPA, 2006.

tinuation of contracts should be highly dependent on reaching pre-defined milestones and deliverables.

The involvement of multiple federal agencies is important, as no single agency is likely to have the needed expertise to address all issues, but it will be important for one agency to take the lead in organizing intra-institutional efforts. Given NCI's current funding level and recent initiatives and interest in biomarker discovery and development, it may seem an obvious choice for the lead agency for this endeavor, but to date it has not yet developed an adequate overarching leadership strategy. The NCI programs described above are relatively narrow in focus, and there is very little coordination or communication among them, with no unifying strategy or oversight. Without appropriate organization and funding, researchers will be unable to muster the resources and knowledge to achieve broad gains, and progress is likely to continue to be slow and piecemeal. Finally, as noted in the last section of the chapter, providing academic scientists with appropriate incentives and resources will be critical if they are to undertake successfully work that has not been traditionally viewed as an academic pursuit.

THE IMPORTANCE OF BIOREPOSITORIES

Analysis of human tissues is essential for biomarker discovery and validation. Human tissue has been collected and stored in biorepositories for more than 100 years in the United States, and it is estimated that there are more than 300 million tissue specimens from more than 175 million cases stored in the United States, with new specimens accumulating at a rate of more than 20 million per year (reviewed by Eiseman and Haga, 1999). These specimens are collected by a broad array of institutions, both federal and private, but most patient samples were originally collected for diagnostic and therapeutic purposes, so the vast majority are not used in research. NIH is the largest single funding source for tissue repositories, and NCI in particular supports many different tissue repositories for research (Box 2-2). However, despite the common funding source, these biobanks

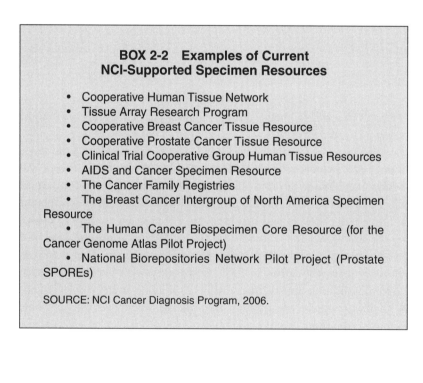

BOX 2-2   Examples of Current NCI-Supported Specimen Resources

- Cooperative Human Tissue Network
- Tissue Array Research Program
- Cooperative Breast Cancer Tissue Resource
- Cooperative Prostate Cancer Tissue Resource
- Clinical Trial Cooperative Group Human Tissue Resources
- AIDS and Cancer Specimen Resource
- The Cancer Family Registries
- The Breast Cancer Intergroup of North America Specimen Resource
- The Human Cancer Biospecimen Core Resource (for the Cancer Genome Atlas Pilot Project)
- National Biorepositories Network Pilot Project (Prostate SPOREs)

SOURCE: NCI Cancer Diagnosis Program, 2006.

vary a great deal in their design, methods, standards, data collection, and informed consent, making it difficult to compare data from different studies or to combine data from different repositories. Indeed, there has been a lack of nationally agreed-on quality control and standard operating procedures, which limits the usefulness of existing collections (Eiseman et al., 2003). In addition, genomic and proteomic analyses require tissue preservation methods that differ from those most commonly used in diagnostic pathology.

In an attempt to address these problems, NCI commissioned a study of 12 U.S. biorepositories (half of which were supported by NIH) to identify best practices, with the goal of developing standard operating procedures. Best practices were identified for all aspects of specimen collection, storage and use, including processing and annotation, storage and distribution, bioinformatics, consumer and user needs, business plan and operations, privacy, ethical and consent issues, intellectual property (IP) and legal issues, and public relations, marketing, and education (Eiseman et al., 2003). NCI also commissioned a blueprint for a National Biospecimen Network (NBN) with the goal of providing a "comprehensive framework for sharing and comparing research results through a robust, flexible, scalable, and secure bioinformatics system that supports the collection, processing, storage, annotation, and distribution of biospecimens and data" (Friede et al., 2003). Primary objectives of the blueprint were to collect biospecimens that were amenable to genomic and proteomic analysis, as well as to ensure uniformity so that data from different studies could be combined or compared. None of the biorepositories examined by Eiseman et al. (2003) had all the characteristics identified as necessary in the NBN report.

The NBN blueprint report has been criticized by some, most notably for its projected costs (Goldberg, 2003). Nonetheless, in 2005, NCI established the Office of Biorepositories and Biospecimen Research with the objective of improving and standardizing biobanking activities and to facilitate the establishment of a National Biospecimen Network (NCI, 2006b). In April 2006, the office put forth first-generation guidelines for all NCI-supported biorepositories (NCI, 2006a), addressing common best practices for research biorepositories, quality assurance, and quality control programs, informatics systems, ways to address ethical, legal, and policy issues (e.g., informed consent, privacy, data security protections, Institutional Review Board oversight, ownership of and access to biospecimens and data), standardized reporting mechanisms, and administration and management structure. Second-generation guidelines, currently being developed in collaboration with the American College of Pathologists and other relevant extramural groups, will propose evidence-based standard operating procedures. A pilot test of the proposed NBN to evaluate the use of best practices for collecting specimens from prostate cancer patients for biomarker research is also ongoing (NCI, 2006c).

The committee supports the development of practice guidelines and standards as well as the harmonization of ethical, legal, and policy issues, and emphasizes the importance of developing strategies to maximize the quality and usefulness of biorepositories while also protecting patient rights. For example, it is important to develop consensus on common data elements for collecting patient information and to make this information and samples easily accessible to researchers. NCI's Early Detection Research Network has made considerable progress in this regard and provides a good model for how to proceed with other biospecimen collections (NCI, 2005, Figure 2-1). Supporting and encouraging the use of electronic patient records would facilitate this work as well.

It is also critical to develop strategies to ensure the confidentiality of identifiable patient health information under the Privacy Rule of the Health Insurance Portability and Accountability Act (HIPAA), without

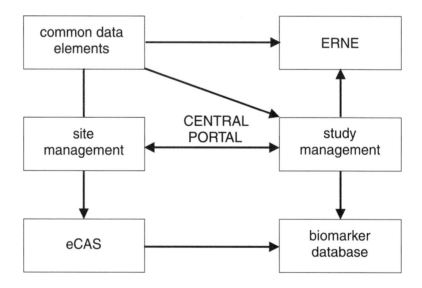

FIGURE 2-1 EDRN informatics infrastructure.
SOURCE: EDRN website, 2006.

impeding research. This may require a reassessment of the privacy regulations established under HIPAA, as well as efforts to promote uniformity in interpretation across states and institutions (Bledsoe, 2004; IOM, 2006b). Promoting interagency harmonization of informed consent is also likely to facilitate research on biomarkers. The committee notes that some privately established biorepositories have been quite successful in dealing with some of the issues that NCI is grappling with, and much could be learned from their example. The Multiple Myeloma Research Consortium Tissue and Data Bank is a prime example (Box 2-3).

However, some challenges remain unaddressed despite these recent activities undertaken by NCI. First, most biomarkers are developed using archived tumor specimens that were collected for other purposes, and many of the discordant findings in the biomarker literature may be due to this retrospective approach (Simon, 2006). Clinical trials to test drugs are usually prospective, with hypotheses, patient selection criteria, analysis plans, and primary endpoints clearly defined in advance of the study. In contrast, biomarker studies are usually performed without a prespecified written protocol defining the hypothesis, eligibility requirements, primary endpoints, or analysis plan. In addition, the specimen population is often very heterogeneous, representing different cancer stages and treatments. This often leads to multiple subset analyses, which increases the chance of false-positive conclusions (Simon, 2005). Many biomarker studies also perform analyses for multiple candidate biomarkers and multiple endpoints, further multiplying the chances for erroneous conclusions (Simon, 2005).

An obvious solution to the problem would be to undertake prospective studies specifically designed to identify and validate predictive or prognostic biomarkers. However, that approach could be prohibitively expensive. A more viable alternative would be to combine prospective therapeutic clinical trials with biomarker studies (Dalton and Friend, 2006), defining appropriate criteria and analyses from the start. NCI should actively encourage and facilitate interaction between biomarker developers and groups involved in clinical research, including therapeutic, screening, prevention, and cohort studies, to enable the prospective collection of high-quality patient samples that are intended to test specific biomarker hypotheses. Open access to all interested biomarker developers (both industry as well as academia) should be a defining feature. True openness will be the best way to ensure these repositories and patient samples are leveraged to the greatest extent possible, and there is good precedent for this from the genome-wide genetic

BOX 2-3   The Multiple Myeloma Research Consortium Tissue and Data Banks

Background and Oversight: Established in 2004, the Multiple Myeloma Research Consortium (MMRC) was founded by Kathy Giusti, the founder and president of the Multiple Myeloma Research Foundation. MMRC is a nonprofit organization created to provide collective, standardized, and technologically integrated resources for academic research. The focus is preclinical, up to and through phase II of clinical trials. MMRC currently has 11 North American academic member institutions and plans to expand membership to European institutions in the future.

Member institutions conduct research in three separate but integrated research cores: genomics, validation, and clinical trials. The MMRC Data Bank integrates laboratory and clinical trial data via secure electronic databases that are accessible to all members. Projects are proposed by member consensus and then sent to a steering committee comprised of the four founding member institutions. All projects are reviewed by a committee composed of representative scientists from each member institution. Projects with a budget over $25,000 are also reviewed by two outside, independent sources. All studies have been prospective, but MMRC plans to conduct retrospective studies in the future, once baseline genomic data have been collected.

Sample Collection and Storage: Bone marrow aspirates and matched blood samples are collected by MMRC's member institutions and by individual donations made through a new "direct-to-patient" program, in which multiple myeloma patients can submit samples taken during clini-

association data now being placed into the public domain with all qualified researchers able to access the raw data and samples for discovery purposes.

NIH should also initiate and sustain funding of biorepositories that are created in conjunction with large cohort studies and clinical trials, and use of these prospectively collected samples should be encouraged for validating biomarkers. When clinical trials end or funding for cohort studies is not renewed, the ability to maintain the biorepositories created in conjunction with the study is often lost (Goodman et al., 2006). The samples collected in these prospective studies may be very valuable for biomarker research

cally necessary bone marrow procurements. All samples are collected under HIPAA consent guidelines and patient samples are assigned a code for each submission in order to protect patient privacy. MMRC currently stores approximately 600 bone marrow aspirate and blood samples in the MMRC Tissue Bank, located at the Mayo Clinic in Scottsdale, Arizona. Samples are collected in accordance with over 50 standard operating procedures developed for collection and handling of samples; in addition, MMRC undergoes internal quality assurance weekly and produces quarterly reports to ensure adherence to good laboratory practice regulations. All samples are annotated with minimal datasets. MMRC is currently transitioning from a paper-based system to an electronic system, managed by LabVantage Solutions. Samples are continuously updated, since they are collected on a rolling basis, as patients enter MMRC-affiliated centers.

Access and Intellectual Property: Currently, only MMRC members may access samples from the tissue bank. However, MMRC is currently working on a $6 million, 3-year genomic sequencing project, the Multiple Myeloma Genomic Initiative, and plans to release all data generated from the project into the public domain. The long-term goal of MMRC is to make all data from every project accessible to the public. Currently, the inventing institution has either sole or joint ownership of intellectual property, but the MMRC retains the right to release data if the principal investigator of a project does not release scientifically sound data in a timely manner.

SOURCES: MMRC, 2006; MMRF, 2006; Young, 2006.

and development, and NIH should consider continued funding for maintenance of biorepositories even if the original study itself is not continued.

In the long term, it may be more feasible to support such biorepositories through public–private consortia (see below). But regardless of how a repository is supported, funding must be sufficient to cover all essential components and activities, including involvement of pathologists to assess sample quality and confirm diagnosis, optimized sample collection and preparation, consistent capture and annotation of clinical patient records, medical informatics and database management, and general administrative and maintenance costs. In addition, extramural experts should be broadly represented on any oversight committees.

THE ROLE OF CONSORTIA

Given the challenge and expense of developing biomarkers, as described above, it may be difficult or impossible for any single company or organization to successfully undertake the work alone. Once validated, biomarkers could prove useful for many stakeholders, including researchers, drug developers, and clinicians, but individual stakeholders may lack the necessary information and resources to effectively develop and validate markers. For example, validating a surrogate endpoint requires costly and lengthy clinical studies. As such, companies inevitably find it cheaper and faster to directly measure the primary endpoint of interest for a particular drug than to first validate the surrogate marker (Fleming, 2005). But once a surrogate marker has been fully validated for a pharmacologic class of treatment regimens, any drug developer could take advantage of the marker to streamline the development of drugs in that class.

Thus, the sharing of precompetitive data and cooperation in developing and validating biomarkers as common goods is paramount to progress in the field. By leveraging the strengths of different partners, consortia offer many advantages over individual efforts (Kettler et al., 2003; Schwartz and Vilquin, 2003; Nishtar, 2004; Chin-Dusting et al., 2005; Croft, 2005). Partnerships can lead to greater efficiency and effectiveness by pooling skills, technologies, and other resources. By sharing costs and risks while also reducing legal and IP barriers, consortia are more likely to take on challenges that, individually, the partners would be unlikely to tackle. Public–private partnerships (PPPs) in particular can more effectively leverage public funding and resources, increase the breadth and depth of representation in science and scientific agendas, and effect a more rapid translation from basic discoveries to public health applications (Mittleman, 2006). Industry, government, and nonprofit organizations all have a potential role to play in such partnerships, and could each make important and unique contributions to the endeavor.

Although private companies normally are inclined to protect their data to maintain a competitive edge, there are numerous precedents of successful PPPs developing tools and generating pre-competitive data to move a field forward, both in biomedical research and in other industries. For example, SEMATECH (Semiconductor Manufacturing Technology), established in

1987, helped U.S. semiconductor suppliers develop new production tools and establish industry-wide consensus on product specifications (Box 2-4). In biomedicine, PPPs constitute a common approach to tropical and neglected diseases (Kettler et al., 2003; Nishtar, 2004; Croft, 2005), and collaborative efforts have already been successful in biomarker development as well. For instance, the HIV Surrogate Markers Collaborative Group, established through multiple partnerships among academia, industry, and government, confirmed the usefulness of HIV RNA as a surrogate marker for testing new anti-HIV drugs. As a result, the Food and Drug Administration (FDA) began to approve drugs based on evidence of lower levels of plasma HIV RNA in response to drug therapy (Behrman, 1999; Mildran, 2006). Also, the International Life Science Institute, and its Health and Environmental Sciences Institute, which bring industry, government agencies, and academics together to share data and information on nutrition, food safety, toxicology, risk assessment, and the environment, has aided the development of biomarker candidates for toxicological assays (both genomic and proteomic).

The SNP Consortium (TSC) is a well-known example of an international public/private collaboration in biomedical research (TSC, 2006; Box 2-5) that demonstrates the willingness of multi-national drug companies to share information to achieve a common goal and shows the feasibility of this collaborative approach for fostering precompetitive work that could benefit the entire field. Single nucleotide polymorphisms (SNPs) are common small DNA variations that occur throughout the human genome. The primary objective of TSC was to create a high-quality, publicly available map of human SNPs, with the hope that it would aid genomic research and the development of genetic-based diagnostics and therapeutics. TSC exceeded its primary goal by identifying and mapping about 10-fold more SNPs than originally planned, while also completing the work in less time and with a smaller budget than had been scheduled. According to Arthur Holden, chief executive officer of TSC, several factors were critical to that accomplishment, including a clear, focused, and unifying objective, a carefully crafted work plan, strong teamwork with experienced management, and preeminent external advisers and investigators (Holden, 2006).

Given the power of partnerships to address unmet needs in biomedicine, a number of new consortia have recently been formed to develop and validate biomarkers (Feigal, 2006; Holden, 2006; IOM, 2006a; Mittleman, 2006). For example, the Critical Path (C-Path) Institute is a publicly funded nonprofit consortium consisting of pharmaceutical industry partners with

**BOX 2-4 SEMATECH: A Successful Public–Private
Partnership in the Semiconductor Industry**

SEMATECH (Semiconductor Manufacturing Technology) was established in 1987, when the Semiconductor Industry Association (SIA) and the Semiconductor Research Corporation (SRC) convened 14 U.S. semiconductor companies to create a nonprofit industry-government consortium aimed at regaining U.S. world leadership of semiconductor manufacturing. Congress hoped that improved semiconductor manufacturing would also bolster the defense technology base and therefore matched industrial funding for SEMATECH by appropriating $100 million annually for five years through DARPA. Members were initially required to contribute 1 percent of their semiconductor sales revenue, with a minimum contribution of $1 million and a maximum of $15 million. SEMATECH's current annual budget is $150 million.

SEMATECH helped U.S. semiconductor suppliers develop next-generation production tools and facilitated manufacturer-supplier communication and collaboration. It also encouraged semiconductor manufacturers to come to consensus about future needs, so that equipment manufacturers were held to just one set of industry specifications rather than different standards for each company. These efforts helped to drastically increase U.S. market share of semiconductor devices. In 1995, SEMATECH announced that it would continue to be funded by industry alone, in order to pursue independent research and development projects, and separate itself from federal objectives. Since then, SEMATECH has developed an international business model and has focused on research and development of new technologies and products. Since 1998, SEMATECH has also fostered collaboration with foreign companies.

SEMATECH has three levels of membership, depending on access to its various programs. Dues for current SEMATECH members are now based on an algorithm dependent on the member's size and annual sales. Intellectual property agreements are covered in participation agreements with each member. IP policies vary by level of membership. However, members of the Advanced Technology Development Facility (ATDF), a subsidiary of SEMATECH, are entitled to 100 percent ownership of IP for the products they develop within ATDF. SEMATECH has had no antitrust litigation to date.

SOURCES: Irwin and Klenlow, 1996; McGowan, 2006; SEMATECH, 2006.

the goal of identifying and validating preclinical and clinical biomarkers for predicting human drug toxicities (Box 2-6). The companies that participate in this consortium share data and methods in order to test and cross-validate one another's markers and methods.

Another new consortium is the Oncology Biomarker Qualification Initiative (OBQI), created to join the efforts of the FDA, NCI, and the Centers for Medicare and Medicaid Services (CMS) with the goal of improving cancer therapeutics and patient outcomes through biomarker development and evaluation. The OBQI aims to facilitate the codevelopment of diagnostic-therapeutic combinations and to reduce the time and cost of drug development by shortening clinical trials through enriched patient populations more likely to respond to therapies. The first project of OBQI will entail a PPP to qualify fluorodeoxyglucose positron emission tomography (FDG-PET) scanning as a marker for drug response in non-Hodgkin's lymphoma.

The Pharmaceutical Biomedical Research Consortium (PBRC) was also recently formed, with the mission of "advancing the field of medicine, through the development and implementation of high quality pre-competitive biomedical consortia developed with its pharmaceutical members, in conjunction with appropriate other partners" (Holden, 2006; Box 2-7). The PBRC intends to undertake a number of independent projects, several of which will focus on biomarkers, including surrogate markers and predictive markers of serious adverse events.

NIH has also recently spearheaded several PPPs to develop biomarkers for common diseases, such as osteoarthritis and Alzheimer's disease, with more projects planned for the future (Mittleman, 2006; NIH, 2006a). A primary goal of NIH is to establish policy regarding use of samples and information from study collections that have already been created. NIH also plans to develop common Institutional Review Boards for multi-centered studies so that a single determination can be provided by a single, common IRB, rather than multiple institutional IRBs providing multiple determinations.

In October 2006, the Foundation for the National Institutes of Health (FNIH), NIH, the FDA, and the Pharmaceutical Research and Manufacturers of America (PhRMA) announced the launch of another public–private biomedical research partnership, the Biomarkers Consortium, to search for and validate new biomarkers. FNIH has already secured $3 million in donations from major pharmaceutical companies for the consortium, and more funders are anticipated to join the effort. Like the OBQI, the first project

BOX 2-5   The SNP Consortium

Origin and Oversight: The SNP Consortium (TSC) was established in April 1999 as a nonprofit organization that provided free and accessible SNP data to researchers and the public in an effort to expedite drug research and discovery. A total of 13 corporations joined the U.K. Wellcome Trust philanthropy as members of TSC to fund SNP research in a collaborative, precompetitive environment. Each member organization was represented on a governing board that was led by an independent chairman. Membership was open to any nonprofit, governmental, or private organization involved in SNP research willing to make a financial contribution equal to other TSC members, although there was a 13-member ceiling for the governing board. The Wellcome Trust pledged $14 million, and each TSC member gave $3 million over the two-year membership term.

SNP identification and analysis was conducted at several affiliated research centers, including the Whitehead Institute, Washington University, the Sanger Center at the Wellcome Genome Research Campus, and the Stanford Genome Center. Data and bioinformatics were managed by Cold Spring Harbor Laboratory.

Data Collection and Release: To ensure representation of the entire population, TSC's SNP research used a pool of DNA samples obtained from 24 individuals from several racial groups. All DNA contributions were anonymous, voluntary, and obtained with informed consent. SNP data were regularly validated by internal quality control assessment and by external auditors; the estimated validation rate of both internal and external analysis was 95 percent.

of this new consortium will be to evaluate the use of FDG-PET to measure response to treatment, but in this case, the group will simultaneously study lung cancer in addition to non-Hodgkin's lymphoma. Other projects under consideration will focus on mental health and diabetes (FNIH, 2006).

All of these new collaborative efforts are commendable, but with the exception of the projects on FDG-PET imaging, none is focused on developing cancer biomarkers. Thus, the committee recommends that industry and other funders of biomedical research establish international public–private consortia to generate and share methods and precompetitive

The purpose of TSC was to maximize the number of SNP discoveries that enter the public domain. Data were released simultaneously to TSC members and the public at approximately quarterly intervals, which allowed for SNP mapping and validation. These data were made available through a consortium website, the dbSNP database, managed by the National Center for Biotechnology Information, and the Human Genome Variation Database (HGVbase). In total, TSC made 12 regular public releases (the final major release in 2001).

TSC set an initial goal of identifying 300,000 and mapping 170,000 SNPs within two years. All SNPs were to be released into the public domain via Internet access. The results of TSC far exceeded the initial goal; by the end of 2001, 1.4 million SNPs were identified, mapped, and released. By March 2005, 2.7 million SNPs had been released to the public, of which approximately 2.3 million were "unencumbered" SNPs, and 2.5 million unique SNPs had been mapped. TSC spent $42 million of available funds, thus remaining under their budget limit. Since completion of TSC's two-year initiative, the discovery phase of SNP identification and mapping is essentially over, and several TSC members have begun researching the frequency of SNPs in certain major world populations as part of the Allele Frequency Project.

Intellectual Property: In order to increase the number of SNPs in the public domain and reduce financial or other IP-related third-party encumbrances to public use, TSC withheld public release of identified SNPs until mapping. Patent applications were filed solely to establish the dates of scientific discoveries of the SNPs mapped.

SOURCE: TSC, 2006.

data on the discovery, validation, and qualification of cancer biomarkers. A more cooperative and comprehensive approach that attempts to leverage and integrate all available data could have an enormous impact on the field. Such efforts could lead to better biomarkers for the entire spectrum of cancer health care, from early detection and disease classification to drug development and treatment planning and monitoring (Bast et al., 2005; Dalton and Friend, 2006). Organizers should examine and learn from past and ongoing biomedical consortia, especially the SNP Consortium, which provides a valuable model. Many complex issues must be addressed

BOX 2-6   Critical Path Institute

The Critical Path Institute (C-Path) was created by the FDA in July 2005 as a nonprofit, publicly funded institute to provide a neutral ground for FDA scientists, academic researchers, and industry to collaborate on and accelerate the development of safe medical products. A major focus is on advancing appropriate applications of predictive safety markers at the regulatory interface, before adverse events happen. There are currently 10 pharmaceutical companies that have signed the C-Path consortial agreement, and 4 others are waiting to join. C-Path administration is comprised of a director and codirector, an advisory committee, and a project manager chosen from the consortium members.

Funding: C-Path has gathered approximately $11 million from states and the cities and counties of Arizona. A basic principle is to obtain public funding for infrastructure and federal appropriations for projects. Although it has no direct funding from drug companies, C-Path will allow industry consortia funding for projects, with FDA oversight.

Member Qualifications: Members must have expertise and programs in safety biomarkers, as well as a willingness to share data, experience, and IP in order to validate products; in other words, they must make IP available to consortium members and enter in consortium

in forming consortia, including governance, data sharing and access, intellectual property management, human subjects protections, and antitrust laws. Lessons from past experience could help to streamline the process and increase the probability of broad participation and success.

DEMONSTRATION PROJECTS TO DEVELOP BIOMARKERS FOR DRUGS ALREADY APPROVED

Most drugs are effective in only a fraction of the patients who receive them (Spear et al., 2001). This is especially true for cancer drugs, with an average drug response rate of less than 25 percent. The variability in response is due largely to the heterogeneity of specific molecular changes in tumors, which cannot be identified by current diagnostic methods. Differences in drug metabolism due to genetic variability (polymorphism) in patients can also contribute to the inconsistency in drug response.

agreements. Members should be willing to commit internal resources to validate other members' safety biomarkers.

Major Foci:
- Validate predictive, preclinical animal model biomarkers to reduce the cost and time of preclinical safety studies
- Provide public access to validated tools
- Provide potential early indicators of clinical safety in drug development and postmarket surveillance
- Provide new tools for FDA to assist in regulatory decision making

Example—Warfarin Pharmacogenetics Project: Improper dosing of warfarin causes unnecessary health care spending and trauma for patients. Genetic variation contributes significantly to dosing variability. The goals of the project are to:
- Investigate how clinical factors and drug interactions affect warfarin response
- Provide an evidence base for labeling
- Aid physicians in determining proper dosing for their patients
- Inform insurers' decisions regarding coverage of genomic tests

SOURCE: Feigal, 2006.

The current approach to cancer treatment is still largely empirical and centered on population-based statistics (Dalton and Friend, 2006). Treatments are assigned according to diagnostic categories that are derived from cancer type and stage, rather than specific molecular changes. Biomarkers that would enable physicians to choose the treatment most likely to benefit a given patient could greatly improve treatment outcome, both in terms of improved effectiveness and in avoiding potentially debilitating but ineffective treatment. These improvements would also enhance the cost-effectiveness of treatment (see Chapter 4 for more information on cost-effectiveness), by increasing the probability that expensive treatments will be effective and by reducing the costs associated with managing toxic side effects.

However, once a drug is approved by the FDA, drug companies have relatively little incentive to develop biomarkers to guide treatment decisions, as this would likely restrict the population of patients treated with

BOX 2-7 Pharmaceutical Biomarker Research Consortium

The Pharmaceutical Biomarker Research Consortium (PBRC) was founded with the goal of accelerating the development and implementation of high-quality, precompetitive biomedical consortia by aggregating industry priorities to develop more efficient and effective research platforms. Platforms are to be standardized and developed in conjunction with pharmaceutical members and other appropriate partners, such as academic researchers and government organizations. Industry members can identify specific projects that they believe could benefit from collaboration rather than independent development. An umbrella legal counsel was set up so that projects could be quickly and efficiently initiated using a nonprofit research consortium.

Key Features:

- No real infrastructure or standing staff—mostly outsourcing and borrowing/sharing
- Biomedical research-focused mission, research focused by-laws and charter
- Pooling of talent, experience, and required specialized consortia skills
- Antitrust protection
- Independent data handling and release
- Execution of desired regulatory standards

Serious Adverse Events Consortium

In April 2006, the FDA approached the PBRC to develop and lead an industry-driven, nonprofit consortium focused on drug-induced serious adverse events (SAEs). Drug-related SAEs are a significant issue

the drug. For example, several new drugs that inhibit the epidermal growth factor receptor (EGFR) were recently approved by the FDA (FDA News, 2003, 2004a, 2004b, 2006). However, in each case, only a small minority of patients with the type of cancer for which the drugs are indicated actually responds to treatment, and to date no biomarkers have been shown to effectively identify patients who will and will not respond to each drug. Furthermore, several additional EFGR inhibitors are in clinical trials, so making the appropriate treatment choice could become even more chal-

for patients, the FDA, industry, and payors. The SAE consortium functions on a one-member, one-vote majority rule and is overseen by an independent chairman and governed by a board of directors.

Purpose: To develop patient-sample networks in order to apply pharmacogenetics to determine the genetic basis of drug-induced SAE and to leverage the resources and talent of large pharmaceutical and biotechnology enterprises, academic researchers, and government in pursuit of that mission.

Intellectual Property: Free and unencumbered markers with equal data access for all parties. Provisional patent applications, which are filed but not reviewed by the U.S. Patent and Trademark Office (PTO), will be used to set a "priority date" or date of invention. Within one year after filing, a provisional patent must be converted to a utility patent application or abandoned. When markers are confirmed and validated, Statutory Invention Registrations (SIRs), which are not examined by the PTO, will be filed. The SIR is a document that permits an inventor to place an invention in the public domain to prevent others from obtaining a patent for it. This is known as a "protective IP strategy."

Specific Goals:
- Develop a coordinated network to support SAE retrospective and prospective discovery and validation of pharmacogenomic markers
- Create a public knowledge base to identify pharmacogenomic markers to predict SAEs
- Apply whole-genome SNP mapping technology to SAE marker development
- Manage IP relating to pharmacogenomic markers useful in predicting SAEs to ensure access for diagnostic and therapeutic applications

SOURCE: Holden, 2006.

lenging if and when these drugs gain FDA approval and enter the market (Grunwald and Hidalgo, 2003; Baker, 2004). Some efforts are under way to identify biomarkers to guide EGFR treatment decisions, but these studies are largely being done in academic settings. Furthermore, these efforts are not coordinated or unified through data sharing or by a common strategy.

Federal agencies and other funders should therefore support demonstration projects to discover and develop biomarkers that can predict the safety and effectiveness of FDA-approved oncology drugs in individual patients, with the goal of selecting appropriate target populations for those drugs. A high-impact finding for a specific disease or drug-targeted pathway would not only improve treatment outcomes for patients, but also could define an optimal approach for the biomarker field and catalyze the diagnostic and pharmaceutical industries to undertake such studies for many cancers and therapeutics by establishing a viable route to market with profitable returns (i.e., viable business models) (IOM, 2006a). Such a finding would also establish precedents for health care providers to use biomarkers to guide therapy decisions and for payors to provide coverage for such tests.

A precedent for using a pharmacogenomic biomarker test to predict the major toxicity of a cancer drug already exists. Irinotecan, used primarily to treat advanced colorectal cancer, is a prodrug that is converted by a cellular enzyme to an active form. This active form of the drug is further modified by a second enzyme known as UGT1A1, allowing it to be eliminated from the body more efficiently (reviewed by Nguyen et al., 2006; Maitland et al., 2006). People with certain polymorphisms in the UGT1A1 gene have reduced ability to clear the active drug from the body and are therefore much more likely to experience dangerous adverse effects from the drug, including severe myelosuppression and diarrhea. As a result of these findings, in 2005 the FDA approved revisions to the safety labeling for irinotecan to recommend reduced dosing in patients who are homozygous for a specific UGT1A1 allele. A month later, the FDA also approved a molecular diagnostic test to identify patients with variations in the UGT1A1 allele that may be at increased risk of adverse side effects from irinotecan.

Another potentially informative case of how pharmacogenomics might be used to predict the effectiveness of a drug is a body of work undertaken to understand the variability in response to the antiestrogen tamoxifen in breast cancer patients (Box 2-8). Estrogen receptor status has been used for many years to identify patients who are likely to respond to tamoxifen, but more recent studies indicate that variations in the CYP2D6 gene might be used to identify nonresponders within that subgroup (Goetz et al., 2005). This could be useful information, as other treatment options, such as aromatase inhibitors, are now available to treat women with ER+ tumors. The investigators were fortunate to have access to a biomarker test already approved by the FDA,[4] so other projects may face additional hurdles in

---

[4]Roche's Amplichip was developed primarily to target drug therapy for a variety of diseases by assessing polymorphisms in the cytochrome p450 gene family, of which CYP2D6 is a member. Other tests for variations in this gene family are also under development.

developing appropriate drug-targeting biomarkers. Nonetheless, this effort provides a working example of what might be possible if a concerted effort is undertaken to identify and validate treatment stratification biomarkers for drugs in use. Questions about how best to conduct such studies will need to be addressed early on. In particular, the studies must be well designed and adequately powered, but since patients are already taking the drugs, it should not be difficult to accrue participants for a study (IOM, 2006a).

THE NEED FOR PATHWAY BIOMARKERS

It is now widely accepted that genetic mutations and epigenetic changes are primary driving forces in the initiation and progression of tumors. Cancers have been increasingly linked to changes that affect how proteins function within signaling pathways that control cell growth and death, motility, metabolism, and genomic integrity (Coleman and Tsongalis, 2006; Esteller, 2006; Varmus, 2006). That is, cancers arise and progress when cell-signaling pathways are altered. Furthermore, much of the heterogeneity among cancer patients can be traced to differences in the specific pathways that have been modified in each tumor. Decades of research have gradually led to the identification and delineation of the pathways that control these vital cell functions, and recent advances in developing molecularly targeted therapies, like imatinib and trastuzumab, are derived from that increased understanding of signaling pathways. In addition, acquired resistance to these cancer drugs is attributed to secondary mutations in critical signaling pathways (Baselga, 2006).

Thus, biomarkers that can detect alterations in specific signaling pathways would be extremely useful for the detection, diagnosis, and treatment of cancers. Classification of tumors by molecular changes rather than by organ site and morphologic appearance could radically change the approach to cancer care. Screening tests could be devised to detect altered pathways common to many cancers, rather than developing many organ-specific tests for cancers. Once cancer is diagnosed, identifying the altered pathways would aid in making individualized treatment decisions. Biomarkers tied to specific drugs, in contrast, can provide only a yes/no answer for that particular drug; they cannot suggest an optimal alternate treatment. In addition, it is widely believed that targeting multiple pathways will be necessary to effectively treat most cancers (Baselga, 2006). Pathway biomarkers would allow for a "systems" approach to diagnosis, treatment, and surveillance (van der Greef and McBurney, 2005), recognizing that pathways operate in the

BOX 2-8   Tamoxifen Therapy and the CYP2D6 Gene

Tamoxifen is highly metabolized by a variety of enzymes to many metabolites with a wide range of potencies. 4-OH-Tam is one of these metabolites that has been studied in vitro for many years because it has a much higher affinity for the estrogen receptors. In recent years, it has become evident that another metabolite (4-hydroxy-N-desmethyl-tamoxifen, known as endoxifen) is equally potent to 4-OH-Tam. Endoxifen is present at about 10-fold higher concentrations than 4-OH-Tam, and the concentration of endoxifen varies a great deal among breast cancer patients. Much of this variability appears to be due to the genetic differences in CYP2D6, the main enzyme that generates endoxifen.

The CYP2D6 gene is known to be highly polymorphic. For example, in Caucasians, about 5–7 percent of the population has no functional enzyme activity (poor metabolizers). This is due to SNPs that cause nonfunctional enzyme activity or, in some cases, the loss of the entire gene. In addition, about 1 percent of the Caucasian population has multiple copies of the CYP2D6 gene and thus elevated metabolic capacity (ultrarapid metabolizers). The frequencies of these poor metabolizers and ultra rapid metabolizers vary greatly among different populations. For example, about 30 percent of Ethiopians are ultrarapid metabolizers, while some populations have much higher rates of poor metabolizers.

Clinical studies have shown that CYP2D6 genetic variations are strong determinants of circulating endoxifen concentrations. Furthermore, a retrospective study of breast tumors from patients taking tamoxifen found that subjects with the CYP2D6*4 allele, which has no functional activity, had more rapid recurrence than patients with wild type

context of interconnected networks. A recent study published in the journal *Nature* described a novel approach to define such pathway signature markers to aid prognosis and predict drug sensitivity (Bild et al., 2006).

Pathway biomarkers could also help identify new drug targets and streamline the drug development process (Stoughton and Friend, 2005; Dalton and Friend, 2006). It can be argued that a lack of financial incentives limits industry investment and efforts to develop narrowly targeted drugs for specific subsets of cancers. However, targeting pathways that are common to many different types of cancer could expand the potential use of and market for new molecularly targeted drugs. For example, pathway biomarkers used to develop drugs for common cancers could potentially

CYP2D6 activity. A prospective study to assess patient outcome when treatment assignment is based on CYP2D6 genotyping has not yet been undertaken, but it is under consideration. For example, postmenopausal patients who are CYP2D6 poor metabolizers could be treated with aromatase inhibitors rather than tamoxifen. The options for premenopausal women are more complicated because aromatase inhibitors are generally not recommended for their treatment, although use in combination with ovarian suppression by ovariectomy or suppression with LHRH agonist could be a possibility.

On October 18, 2006, the FDA's Clinical Pharmacology Subcommittee of the Advisory Committee for Pharmaceutical Science recommended that the FDA revise tamoxifen's drug label to include a warning that postmenopausal women who are CYP2D6 poor metabolizers and are taking tamoxifen to treat breast cancer have an increased risk for breast cancer recurrence. The subcommittee also recommended that the label should include a warning that certain antidepressants may reduce tamoxifen's effectiveness. The subcommittee panel members did not come to a consensus about whether the FDA should recommend CYP2D6 genetic testing for tamoxifen patients, but the majority was in favor of including it as an option in the appropriate section of the drug packaging insert. The panel's recommendations do not require the FDA to make changes to the tamoxifen label; however, the FDA usually does follow the advice of subcommittees.

SOURCES: Lee et al., 2003; Stearns et al., 2003; Desta et al., 2004; Johnson et al., 2004; Gjerde et al., 2005; Jin et al., 2005; Lim et al., 2005; AJHP News, 2006; Bernard et al., 2006; FDA, 2006; IOM, 2006; Knox et al., 2006; Skaar, 2006.

---

also be useful in rarer forms of cancer that are more difficult to study and would offer smaller returns to developers.

This emphasis on pathways could also invigorate the field of biomarker development itself. Biomarkers that are exclusively focused on a particular drug must be developed in conjunction with each new drug, at a high cost and with considerable risk. If an experimental drug does not achieve FDA approval, work on the associated biomarker would be for naught (IOM, 2006a). And even if FDA approval is obtained, the biomarker could still become obsolete if newer, more effective drugs become available and therapy guidelines change. In contrast, pathway markers are more likely to be applicable to the development of any new drug that targets an essential pathway. Broad applicability would be optimized by emphasizing the development of objectively quantifiable biomarkers, rather than qualitative or semi-quantitative assays such as immunohistochemistry. This broader applicability will increase the potential market and also reduce the risk associated with the development process, thus improving the odds of profitability for the company. One caveat is that the requirements for sensitivity, specificity, precision, and accuracy of a biomarker may vary among different diseases, so markers would not necessarily be directly transferable. But the secondary development process for another disease is likely to be shorter and less expensive than starting from scratch.

THE NEED FOR SUPPORT OF
TRANSLATIONAL RESEARCH ACTIVITIES

NIH and NCI have both recently stressed the need to support and facilitate translational research to ensure that critical basic science discoveries move "from bench to bedside" and that unmet medical needs in turn drive further bench research. The NIH Roadmap noted that "growing barriers between clinical and basic research, along with the ever increasing complexities involved in conducting clinical research, are making it more difficult to translate new knowledge to the clinic" (NIH, 2006b). An annual report from the President's Cancer Panel concluded that "the translational research infrastructure is inadequate to enable the work that needs to be done; resources must be committed to develop the tools and workforce required. Increased funding for translation-oriented research—particularly collaborative, team efforts—is urgently needed across the translation continuum. Targeted Federal funding for translation-oriented research is drastically out of balance relative to financial commitments to basic science. Ways must be found to increase human tissue and clinical research resources without slowing the discovery engine. Supplemental funding may offer a temporary solution, but will be inadequate in the long term" (President's Cancer Panel, 2005). Similarly, the NCI's Translational Research Working Group (TRWG), recently appointed to evaluate the status of NCI's investment in translational research and to envision the future, concluded that translational research is not well coordinated across NCI and that the resulting fragmented efforts are often duplicative and could lead to missed opportunities (Goldberg, 2006; NCI, 2006d). Specifically, a draft report from the TRWG concluded that

- the absence of clearly designated funding and adequate incentives for researchers threatens the perceived importance of translational research in NCI;
  - the absence of a structured, consistent review and prioritization process tailored to the characteristics and goals of translational research makes it difficult to direct resources to critical needs and opportunities;
  - translational research core activities are often duplicative and inconsistently standardized, with capacity poorly matched to need;
  - the multidisciplinary nature of translational research and the need to integrate sequential steps in complex development pathways warrants dedicated project management resources; and
  - insufficient collaboration and communication between basic and clinical scientists, and the paucity of effective training opportunities limits the supply of experienced translational researchers.

Support for translational research activities will be critical for developing and validating putative biomarkers. Initial discoveries of potential biomarkers are often published in high-impact journals, but subsequent work to confirm and validate those findings often does not merit publication in those same journals. Furthermore, such validation work often takes many years to complete and can require an interdisciplinary team approach to science that is not the norm in academia (reviewed by IOM, 2003; Gray, 2006; Kaiser, 2006a). The academic culture traditionally has not been supportive of faculty that engage in team science or translational research; promotion and reward structures are designed to recognize individual initiative and accomplishment. Thus, it will be important to consider how academic organizational structures, metrics for academic promotion, and the cultures of biomedical research can better support team building and multidisciplinary science. Key factors for success will include providing sufficient time, resources, and rewards for faculty who undertake translational research (Gray, 2006). Training programs that specifically deal with the many complexities of this work are also needed to help new translational investigators get started and become established (Kaiser, 2006a).

Although not a traditional NIH funding focus, several recent initiatives have been undertaken to foster translational research. For example, NIH's new Clinical and Translational Science Award Program encourages institutions to develop new approaches to clinical and translation research, including new organizational models and training programs, and to develop novel clinical research methodologies (DHHS, 2006a; Gray, 2006; Kaiser,

2006b). Through this program, 12 institutions recently received 5-year awards totaling $108 million for the first year. The program is intended to eventually replace the 50-year old NIH program of General Clinical Research Centers, which currently consists of approximately 60 facilities with beds for patients participating in clinical studies. The NCI TRWG draft report also recommends a new organizational structure to coordinate NCI's translational research, with designated leadership and budget and oversight by an external advisory committee. Some private funders, such as the Howard Hughes Medical Institute, the Burroughs Wellcome Fund, and the Doris Duke Charitable Foundation, have put a recent emphasis on translational research as well (Kaiser, 2006a).

Nonetheless, there is concern that funding and other support will not be maintained well enough to sustain a nascent, growing field of translational specialists (Kaiser, 2006a). But continued funding from federal and private sponsors of this work is essential for progress in reaching the goal of personalized medicine. Indeed, NCI has noted that the development of new diagnostic tests, cancer treatments, and other interventions that benefit people with cancer and people at risk for cancer will rely on strong translational research collaborations between basic and clinical scientists to generate novel approaches (NCI, 2006d).

SUMMARY AND CONCLUSIONS

The discovery and development of biomarkers entails a complex, multistage process, with many challenges that must be overcome to make meaningful progress in the field. Despite a few spectacular successes, the number of biomarkers used in drug development or clinical practice is very small, and most putative biomarkers never advance beyond the discovery stage. Moreover, the limitations of current technology render many discovery efforts inefficient and inadequate. Changes are needed to streamline the process and make the most of limited resources available for biomarker research and development.

First, a more organized, comprehensive approach to biomarker discovery is needed. Such an approach would more effectively foster technological innovation and could lead to more efficient, systematic searches for potential biomarkers. Second, international public–private consortia are needed to generate and share methods and precompetitive data on the validation and qualification of cancer biomarkers. Given the accomplishments of previous endeavors like The SNP Consortium, such collaborations are likely to reduce the cost and risk of biomarker development and enable the field to move forward more efficiently.

Funders of biomedical research should also place more emphasis on developing pathway biomarkers and on developing biomarkers for drugs already in use. A focus on quantifiable biomarkers of signaling pathways rather than individual cancers or drugs could increase the applicability of biomarkers and thus increase the potential for return on investments by sponsoring companies.

Demonstration projects to develop biomarker tests that could determine which patients are most likely to benefit from drugs that are already in the clinic would not only improve treatment outcomes for patients, but also could catalyze industry and academia to undertake such studies by establishing a viable route to market and by delineating a viable business strategy.

Ensuring the availability of high-quality and well-annotated patient samples that have been collected in prospective studies will be crucial to progress in discovering and developing biomarkers. Thus, funders of biomedical research funders should initiate and sustain funding for biorepositories of such patient samples collected in conjunction with large cohort studies and clinical trials, and use of these samples should be encouraged for validating biomarkers. NCI in particular should actively encourage and facilitate interaction between biomarker developers and clinical trials groups to enable this prospective collection of patient samples.

Collectively, these strategies could lead to better biomarkers for the entire spectrum of cancer health care, from early detection and disease classification to drug development and treatment planning and monitoring, and they could bring personalized medicine closer to being a reality.

REFERENCES

Aebersold R, Anderson L, Caprioli R, Druker B, Hartwell L, Smith R. 2005. Perspective: A program to improve protein biomarker discovery for cancer. *Journal of Proteome Research* 4(4):1104-1109.

AJHP (American Journal of Health-System Pharmacy) News. 2006. *Genetics examined in tamoxifen's effectiveness.* [Online]. Available: http://www.ashp.org/news/ShowArticle. cfm?id=17501 [accessed November 2006].

Altman DG, Riley RD. 2005. Primer: An evidence-based approach to prognostic markers. *Nature Clinical Practice Oncology* 2(9):466-472.

Anderson NL, Anderson NG. 2002. The human plasma proteome: History, character, and diagnostic prospects. *Molecular and Cellular Proteomics* 1(11):845-867.

Azzazy HM, Mansour MM, Kazmierczak SC. 2006. Nanodiagnostics: A new frontier for clinical laboratory medicine. *Clinical Chemistry* 52(7):1238-1246.

Baker M. 2004. EGFR inhibitors square off at ASCO. *Nature Biotechnology* 22(6):641.

Baker SG, Kramer BS, Prorok PC. 2004. Development tracks for cancer prevention markers. *Disease Markers* 20(2):97-102.

Baselga J. 2006. Targeting tyrosine kinases in cancer: The second wave. *Science* 312(5777):1175-1178.

Bast RC Jr, Lilja H, Urban N, Rimm DL, Fritsche H, Gray J, Veltri R, Klee G, Allen A, Kim N, Gutman S, Rubin MA, Hruszkewycz A. 2005. Translational crossroads for biomarkers. *Clinical Cancer Research* 11(17):6103-6108.

Behrman RE. 1999. FDA approval of antiretroviral agents: An evolving paradigm. *Drug Information Journal* (33):337-341.

Bernard S, Neville KA, Nguyen AT, Flockhart DA. 2006. Interethnic differences in genetic polymorphisms of CYP2D6 in the U.S. population: Clinical implications. *The Oncologist* 11(2):126-135.

Bild AH, Yao G, Chang JT, Wang Q, Potti A, Chasse D, Joshi MB, Harpole D, Lancaster JM, Berchuck A, Olson JA Jr, Marks JR, Dressman HK, West M, Nevins JR. 2006. Oncogenic pathway signatures in human cancers as a guide to targeted therapies. *Nature* 439(7074):353-357.

Bledsoe M. 2004. HIPAA models for repositories. *ISBER (International Society for Biological and Environmental Repositories) Newsletter* 4(1):3-8.

Chin-Dusting J, Mizrahi J, Jennings G, Fitzgerald D. 2005. Outlook: Finding improved medicines: The role of academic-industrial collaboration. *Nature Reviews Drug Discovery* 4(11):891-897.

Citron ML. 2004. Dose density in adjuvant chemotherapy for breast cancer. *Cancer Investigation* 22(4):555-568.

Coleman WB, Tsongalis GJ. 2006. Molecular mechanisms of human carcinogenesis. *EXS* (96):321-349.

Cottingham K. 2006. Speeding up biomarker discovery. *Journal of Proteome Research* 5(5):1047-1048.

Cristoni S, Bernardi LR. 2004. Bioinformatics in mass spectrometry data analysis for proteomics studies. *Expert Rev Proteomics* 1(4):469-483.

Croft SL. 2005. Public–private partnership: From there to here. *Transactions of the Royal Society of Tropical Medicine and Hygiene* 99 Suppl 1:S9-S14.

Dalton WS, Friend SH. 2006. Cancer biomarkers—an invitation to the table. *Science* 312(5777):1165-1168.

DARPA (Defense Advanced Research Projects Agency). 2006. *DARPA Webpage*. [Online]. Available: http://www.darpa.mil/ [accessed July 2006].

De Bortoli M, Biglia N. 2006. Gene expression profiling with DNA microarrays: Revolutionary tools to help diagnosis, prognosis, treatment guidance, and drug discovery. In: Gasparini G, Hayes DF, *Biomarkers in Breast Cancer: Molecular Diagnostics for Predicting and Monitoring Therapeutic Effect*. Totowa, NJ: Humana Press Inc. Pp. 47-61.

de Hoog CL, Mann M. 2004. Proteomics. *Annual Review of Genomics and Human Genetics* 5:267-293.

Desta Z, Ward BA, Soukhova NV, Flockhart DA. 2004. Comprehensive evaluation of tamoxifen sequential biotransformation by the human cytochrome P450 system in vitro: Prominent roles for CYP3A and CYP2D6. *Journal of Pharmacology and Experimental Therapeutics* 310(3):1062-1075.

DHHS (Department of Health and Human Services). 2005. December. *Advanced Proteomic Platforms and Computational Sciences for the NCI Clinical Proteomic Technologies Initiative.* [Online]. Available: http://grants.nih.gov/grants/guide/rfa-files/rfa-ca-07-005.html [accessed July 11, 2006].

———. 2006a. *Institutional Clinical and Translational Science Award.* [Online]. Available: http://grants.nih.gov/grants/guide/rfa-files/RFA-RM-06-002.html [accessed July 2006].

———. 2006b February. *Clinical Proteomic Technology Assessment for Cancer.* [Online]. Available: http://grants.nih.gov/grants/guide/rfa-files/RFA-CA-07-012.html [accessed July 11, 2006].

Diamandis EP. 2004. Mass spectrometry as a diagnostic and a cancer biomarker discovery tool: Opportunities and potential limitations. *Molecular and Celullar Proteomics* 3(4):367-378.

EDRN website. 2006. [Online]. Available: http://edrn.nci.nih.gov [accessed October 2006].

Eiseman E, Haga, S. 1999. *Handbook of Human Tissue Sources.* Washington, DC: RAND.

Eiseman E, Bloom G, Brower J, Clancy N, Olmsted S. 2003. *Case Studies of Existing Human Tissue Repositories.* Arlington, VA: RAND Corporation.

Englbrecht CC, Facius A. 2005. Bioinformatics challenges in proteomics. *Combinatorial Chemistry and High Throughput Screening* 8(8):705-715.

Esteller M. 2006. Epigenetics provides a new generation of oncogenes and tumour-suppressor genes. *British Journal of Cancer* 94(2):179-183.

Fan TW, Lane AN, Higashi RM. 2004. The promise of metabolomics in cancer molecular therapeutics. *Current Opinion in Molecular Therapeutics* 6(6):584-592.

FDA (Food and Drug Administration). October 18, 2006. *Summary Minutes of the Advisory Committee Pharmaceutical Science Clinical Pharmacology Subcommittee October 18-19, 2006.* [Online]. Available: http://www.fda.gov/OHRMS/DOCKETS/AC/06/minutes/2006-4248m1.pdf [accessed November 2006].

FDA News. May 5, 2003. *FDA Approves New Type of Drug for Lung Cancer.* [Online]. Available: http://www.fda.gov/bbs/topics/NEWS/2003/NEW00901.html [accessed September 2006].

———. February 12, 2004a. *FDA Approves Erbitux for Colorectal Cancer.* [Online]. Available: http://www.fda.gov/bbs/topics/NEWS/2004/NEW01024.html [accessed September 2006].

———. November 19, 2004b. *FDA Approves New Drug for the Most Common Type of Lung Cancer.* [Online]. Available: http://www.fda.gov/bbs/topics/news/2004/NEW01139.html [accessed September 2006].

———. September 27, 2006. *FDA Approves New Drug for Colorectal Cancer, Vectibix.* [Online]. Available: http://www.fda.gov/bbs/topics/NEWS/2006/NEW01468.html [accessed October 19, 2006].

Feigal E. 2006. *Partnerships to Accelerate Innovation.* Presentation at the meeting of the National Cancer Policy Forum, Washington, DC.

Feng Z, Prentice R, Srivastava S. 2004. Research issues and strategies for genomic and proteomic biomarker discovery and validation: A statistical perspective. *Pharmacogenomics* 5(6):709-719.

Fleming TR. 2005. Surrogate endpoints and FDA's accelerated approval process. *Health Affairs* 24(1):67-78.

FNIH (Foundation of the National Institutes of Health). 2006. Public–private partnership forms The Biomarkers Consortium to advance the science of personalized medicine: Lung cancer and lymphoma set for initial investigations, prospective projects in major depression and diabetes cited. Bethesda, MD: Foundation of the National Institutes of Health.

Friede A, Grossman R, Hunt R, Li R, Stern S (editors). September 2003. *National Biospecimen Network Blueprint.* Durham, NC: Constella Group, Inc.

Gjerde J, Hauglid M, Breilid H, Lundren S, Varhaug JE, Kinsanga ER, Mellgren G, Steen VM, Lien EA. 2005. Relationship between CYP2D6 and SULT1A1 genotypes and serum concentrations of tamoxifen and its metabolites during steady state treatment in breast cancer patients. Abstract. *San Antonio Breast Cancer Symposium* (June):5086.

Goetz MP, Rae JM, Suman VJ, Safgren SL, Ames MM, Visscher DW, Reynolds C, Couch FJ, Lingle WL, Flockhart DA, Desta Z, Perez EA, Ingle JN. 2005. Pharmacogenetics of tamoxifen biotransformation is associated with clinical outcomes of efficacy and hot flashes. *Journal of Clinical Oncology* 23(36):9312-9318.

Goldberg K. 2006. NCI's translational research uncoordinated, advisors say. *The Cancer Letter* 32(24):4-5.

Goldberg P. 2003. A tissue bank to break the bank? NCI, Dialogue plan expensive resource. *The Cancer Letter* 29(32):1-8.

Goodman GE, Thornquist MD, Edelstein C, Omenn GS. 2006. Biorepositories: Let's not lose what we have so carefully gathered! *Cancer Epidemiology, Biomarkers & Prevention* 15(4):599-601.

Gray M. 2006. Broad cultural shift needed to maximize potential of translational research. *Research Policy Alert.* [Online]. Available: http://www.aimbe.org/assets/library/66_012406.pdf [accessed August 2006].

Grunwald V, Hidalgo M. 2003. Developing inhibitors of the epidermal growth factor receptor for cancer treatment. *Journal of the National Cancer Institute* 95(12):851-867.

Hartwell L. 2005. How to build a cancer sensor system. *The Scientist* 18-19.

Hayes DF, Trock B, Harris AL. 1998. Assessing the clinical impact of prognostic factors: When is "statistically significant" clinically useful? *Breast Cancer Research and Treatment* 52(1-3):305-319.

Holden A. 2006. *IOM Session on Biomarker Consortia.* Presentation at the meeting of the National Cancer Policy Forum, Washington, DC.

Hudis CA. 2005. Clinical implications of antiangiogenic therapies. *Oncology (Williston Park)* 19(4 Suppl 3):26-31.

IOM (Institute of Medicine). 2003. *Large-Scale Biomedical Science: Exploring Strategies for Future Research.* Nass S, Stillman B, eds. Washington, DC: The National Academies Press.

———. 2006a. *Developing Biomarker-Based Tools for Cancer Screening, Diagnosis, and Treatment: The State of the Science, Evaluation, Implementation, and Economics. A Workshop.* Patlak M, Nass S, rapporteurs. Washington, DC: The National Academy Press.

———. 2006b. *Effects on Health Research of the HIPAA Privacy Rule: A Workshop.* National Cancer Policy Forum, Herdman R, Moses H, rapporteurs. Washington, DC: The National Academies Press.

Irwin D, Klenlow P. 1996, November. *SEMATECH: Purpose and Performance.* Presentation at the meeting of the Proceedings of the National Academy of Sciences, Washington, DC. National Academy Press.

Jacobs JM, Adkins JN, Qian WJ, Liu T, Shen Y, Camp DG 2nd, Smith RD. 2005. Utilizing human blood plasma for proteomic biomarker discovery. *Journal of Proteome Research* 4(4):1073-1085.

Jin Y, Desta Z, Stearns V, Ward B, Ho H, Lee KH, Skaar T, Storniolo AM, Li L, Araba A, Blanchard R, Nguyen A, Ullmer L, Hayden J, Lemler S, Weinshilboum RM, Rae JM, Hayes DF, Flockhart DA. 2005. CYP2D6 genotype, antidepressant use, and tamoxifen metabolism during adjuvant breast cancer treatment. *Journal of the National Cancer Institute* 97(1):30-39.

Johnson MD, Zuo H, Lee KH, Trebley JP, Rae JM, Weatherman RV, Desta Z, Flockhart DA, Skaar TC. 2004. Pharmacological characterization of 4-hydroxy-N-desmethyl tamoxifen, a novel active metabolite of tamoxifen. *Breast Cancer Research and Treatment* 85(2):151-159.

Kaiser J. 2006a. A cure for medicine's ailments? *Science* 311:1852-1854.

Kaiser J. 2006b. NIH funds a dozen "homes" for translational research. *Science* 314(5797):237.

Kelloff GJ, Lippman SM, Dannenberg AJ, Sigman CC, Pearce HL, Reid BJ, Szabo E, Jordan VC, Spitz MR, Mills GB, Papadimitrakopoulou VA, Lotan R, Aggarwal BB, Bresalier RS, Kim J, Arun B, Lu KH, Thomas ME, Rhodes HE, Brewer MA, Follen M, Shin DM, Parnes HL, Siegfried JM, Evans AA, Blot WJ, Chow WH, Blount PL, Maley CC, Wang KK, Lam S, Lee JJ, Dubinett SM, Engstrom PF, Meyskens FL Jr, O'Shaughnessy J, Hawk ET, Levin B, Nelson WG, Hong WK. 2006. Progress in chemoprevention drug development: The promise of molecular biomarkers for prevention of intraepithelial neoplasia and cancer—A plan to move forward. *Clinical Cancer Research* 12(12):3661-3697.

Kettler H, White K, Jordan S. 2003. *Valuing Industry Contributions to Public–Private Partnerships for Health Product Development.* Geneva, Switzerland: The Initiative on Public–Private Partnerships for Health, Global Forum for Health Research.

Kiviat NB, Critchlow CW. 2002. Novel approaches to identification of biomarkers for detection of early stage cancer. *Disease Markers* 18(2):73-81.

Knox SK, Ingle JN, Suman VJ, Rae JM, Flockhart DA, Zeruesenay D, Ames MM, Visscher DW, Perez EA, Goetz MP. 2006. Cytochrom p450 2D6 status predicts breast cancer relapse in women receiving adjuvant tamoxifen. *Journal of Clinical Oncology* 24(18S)(June 20 Supplement):504.

Kolch W, Mischak H, Pitt AR. 2005. The molecular make-up of a tumour: Proteomics in cancer research. *Clinical Science (London, England)* 108(5):369-383.

LaBaer J. 2005. So, you want to look for biomarkers (Introduction to the special biomarkers issue). *Journal of Proteome Research* 4(4):1053-1059.

Lee KH, Ward BA, Desta Z, Flockhart DA, Jones DR. 2003. Quantification of tamoxifen and three metabolites in plasma by high-performance liquid chromatography with fluorescence detection: application to a clinical trial. *Journal of Chromatography B, Analytical technologies in the biomedical and life sciences* 791(1-2):245-253.

Lim YC, Desta Z, Flockhart DA, Skaar TC. 2005. Endoxifen (4-hydroxy-N-desmethyl-tamoxifen) has anti-estrogenic effects in breast cancer cells with potency similar to 4-hydroxy-tamoxifen. *Cancer Chemotherapy and Pharmacology* 55(5):471-478.

Maitland ML, Vasisht K, Ratain MJ. 2006. TPMT, UGT1A1 and DPYD: Genotyping to ensure safer cancer therapy? *Trends in Pharmacological Sciences* 27(8):432-437.

McGowan D. 2006. SEMATECH Media Relations. Personal communication. May 31, 2006.

Mildran D. 2006. Former Co-Chair, HIV Surrogate Markers Collaborative Group. Personal communication. June 2006.

Mittleman B. 2006, June. *Biomarkers and Partnerships.* Presentation at the meeting of the National Cancer Policy Forum, Washington, DC.

MMRC (Multiple Myeloma Research Consortium). 2006. *Welcome to the MMRC.* [Online]. Available: http://www.themmrc.org [accessed June 2006].

MMRF (Multiple Myeloma Research Foundation). 2006. *Multiple Myeloma Research Foundation.* [Online]. Available: http://www.multiplemyeloma.org [accessed June 2006].

NCI (National Cancer Institute). 1999. *Unconventional Innovations Program (UIP).* [Online]. Available: http://otir.cancer.gov/programs/ati_uip.asp [accessed November, 2006].

———. September 2004. National Cancer Institute Announces Major Commitment to Nanotechnology for Cancer Research. [Online]. http://www.nci.nih.gov/newscenter/pressreleases/nanotechPressRelease/print?page=&keyword= [accessed November 2006].

———. March 2005. *The Early Detection Research Network.* DHHS. [Online] [accessed July 2006].

———. 2006a. *OBBR First-Generation Guidelines for NCI-Supported Biorepositories.* [Online]. Available: http://biospecimens.cancer.gov/biorepositories/guidelines_full_formatted. asp [accessed July 11, 2006].

———. 2006b. *OBBR (Office of Biorepositories and Biospecimen Research) Mission and Goals.* [Online]. Available: http://biospecimens.cancer.gov/biorepositories/obbr_mission.asp [accessed July 11, 2006].

———. 2006c. *Prostate SPORE National Biospecimen Network Pilot.* [Online]. Available: http://prostatebnpilot.nci.nih.gov/ [accessed July 11, 2006].

———. 2006d. *Translational Research Working Group.* [Online]. Available: http://www.cancer.gov/trwg/ [accessed June 21,2006].

NCI Cancer Diagnosis Program. 2006. *NCI-Supported Specimen Resources.* [Online]. Available: http://www.cancerdiagnosis.nci.nih.gov/specimens/finding.html#resources [accessed July 2006].

Nguyen H, Tran A, Lipkin S, Fruehauf JP. 2006. Pharmacogenomics of colorectal cancer prevention and treatment. *Cancer Investigation* 24(6):630-639.

NIGMS (National Institute of General Medical Sciences). 2006. *Protein Structure Initiative.* [Online]. Available: http://www.nigms.nih.gov/Initiatives/PSI [accessed July 2006].

NIH (National Institutes of Health). 2006a. *NIH Public-Private Partnerships.* [Online]. Available: http://ppp.od.nih.gov/ [accessed June 2006].

———. 2006b. *NIH Roadmap for Medical Research: Re-engineering the Clinical Research Enterprise.* [Online]. Available: http://nihroadmap.nih.gov/clinicalresearch/overview-translational.asp [accessed July 2, 2006].

Nishtar S. 2004. Public–private "partnerships" in health—A global call to action. *Health Research Policy and Systems* 2(1):5.

Norton L. 1997. Evolving concepts in the systemic drug therapy of breast cancer. *Seminars in Oncology* 24(4 Suppl 10):S10-3-S10-10.

Norvell J, Berg J. 2005. The protein structure initiative, five years later. *The Scientist* 19(20):30.

NSABP (National Surgical Adjuvant Breast and Bowel Project). 2006. *Study of Tamoxifen and Raloxifene.* [Online]. Available: http://www.nsabp.pitt.edu/star/index.asp [accessed July 6, 2006].

Omenn GW, States DJ, Adamski M, Blackwell TW, Menon R, Hermjakob H, Apweiler R, Haab BB, Simpson RJ, Eddes JS, Kapp EA, Moritz RL, Chan DW, Rai AJ, Admon A, Aebersold R, Eng J, Hancock WS, Hefta S, Meyer H, Paik Y, Yoo J, Ping P, Pounds J, Adkins J, Qian X, Wang R, Wasinger V, Wu CY, Zhao X, Zeng R, Archakov A, Tsugita A, Beer I, Pandey A, Pisano M, Andrews P, Tammen H, Speicher DW, Hanash SM. 2005. Overview of the HUPO Plasma Proteome Project: results from the pilot phase with 35 collaborating laboratories and multiple analytical groups, generating a core dataset of 3020 proteins and a publicly-available database. *Proteomics* 5:3226-3245.

Ong S, Mann M. 2005. Mass spectrometry-based proteomics turns quantitative. *Nature Chemical Biology* 1:252-262.

Pepe MS, Etzioni R, Feng Z, Potter JD, Thompson ML, Thornquist M, Winget M, Yasui Y. 2001. Phases of biomarker development for early detection of cancer. *Journal of the National Cancer Institute* 93(14):1054-1061.

President's Cancer Panel. June 2005. *Translating Research into Cancer Care: Delivering on the Promise.* [Online]. Available: http://deainfo.nci.nih.gov/ADVISORY/pcp/pcp04-05rpt/ReportTrans.pdf [accessed July 10, 2006].

Quackenbush J. 2006. Microarray analysis and tumor classification. *New England Journal of Medicine* 354(23):2463-2472.

Ransohoff DF. 2004. Rules of evidence for cancer molecular-marker discovery and validation. *Nature Reviews Cancer* 4(4):309-314.

———. 2005. Bias as a threat to the validity of cancer molecular-marker research. *Nature Reviews Cancer* 5(2):142-149.

Ross JS, Gray K, Mosher R, Stec J. 2004. Molecular techniques in cancer diagnosis and management. In: Nakamura RM, Grody WW, Wu JT, Nagle RB, *Cancer Diagnostics: Current and Future Trends.* Totowa, NJ: Humana Press, Inc. Pp. 325-360.

Schwartz K, Vilquin JT. 2003. Building the translational highway: Toward new partnerships between academia and the private sector. *Nature Medicine* 9(5):493-495.

SEMATECH. 2006. *SEMATECH webpage.* [Online] Available: http://www.sematech.org/ [accessed June 2006].

Simon R. 2005. Roadmap for developing and validating therapeutically relevant genomic classifiers. *Journal of Clinical Oncology* 23(29):7332-7341.

———. 2006. Guidelines for the design of clinical studies for the development and validation of therapeutically relevant biomarkers and biomarker-based classification systems. In: Gasparini G, Hayes D, *Biomarkers in Breast Cancer: Molecular Diagnostics for Predicting and Monitoring Therapeutic Effect.* Totowa, NJ: Humana Press, Inc. Pp. 3-15.

Skaar, T. 2006. Associate Director, General Clinical Research Center Pharmacogenetics Core Laboratory, Indiana University. Personal communication.

Spear BB, Heath-Chiozzi M, Huff J. 2001. Clinical application of pharmacogenetics. *Trends in Molecular Medicine* 7(5):201-204.

Srivastava S, Wagner J. 2002. Surrogate endpoints in medicine. *Disease Markers* 18:39-40.

Stearns V, Johnson MD, Rae JM, Morocho A, Novielli A, Bhargava P, Hayes DF, Desta Z, Flockhart DA. 2003. Active tamoxifen metabolite plasma concentrations after coadministration of tamoxifen and the selective serotonin reuptake inhibitor paroxetine. *Journal of the National Cancer Institute* 95(23):1758-1764.

Stoughton RB, Friend SH. 2005. How molecular profiling could revolutionize drug discovery. *Nature Reviews Drug Discovery* 4(4):345-350.

Tinker AV, Boussioutas A, Bowtell DD. 2006. The challenges of gene expression microarrays for the study of human cancer. *Cancer Cell* 9(5):333-339.

TSC (The SNP Consortium). 2006. *The SNP Consortium Ltd.* [Online]. Available: http://snp.cshl.org [accessed June 2006].

Tyers M, Mann M. 2003. From genomics to proteomics. *Nature* 422(6928):193-197.

van der Greef J, McBurney RN. 2005. Innovation: Rescuing drug discovery: in vivo systems pathology and systems pharmacology. *Nature Reviews Drug Discovery* 4(12):961-967.

Varmus H. 2006. The new era in cancer research. *Science* 312(5777):1162-1165.

Wagner JA. 2002. Overview of biomarkers and surrogate endpoints in drug development. *Disease Markers* 18(2):41-46.

Wagner, JA. 2006, March 20. *Biomarker Validation and Qualification: Fitness for Use.* Presentation at the IOM workshop on Developing Biomarker-based Tools for Cancer Screening, Diagnosis, and Treatment. Washington, DC.

Weckwerth W, Morgenthal K. 2005. Metabolomics: from pattern recognition to biological interpretation. *Drug Discovery Today* 10(22):1551-1558.

Yezhelyev MV, Gao X, Xing Y, Al-Hajj A, Nie S, O'Regan RM. 2006. Emerging use of nanoparticles in diagnosis and treatment of breast cancer. *Lancet Oncology* 7(8):657-667.

Young S. 2006. MMRC Executive Director. Personal Communication. June 7, 2006.

Zolg JW, Langen H. 2004. How industry is approaching the search for new diagnostic markers and biomarkers. *Molecular and Cellular Proteomics* 3(4):345-354.

3

Guidelines, Standards, Oversight, and Incentives Needed for Biomarker Development

REVIEW OF CURRENT FDA OVERSIGHT FOR BIOMARKER TESTS

Since the passage of the Food, Drug and Cosmetic Act in 1938, the safety and effectiveness of medical diagnostics has been overseen by the Food and Drug Administration (FDA, 2006c). More specifically, the FDA has regulatory jurisdiction over any device or in vitro reagent that is "intended for use in the diagnosis of disease or other conditions, or in the cure, mitigation, treatment or prevention of disease, in man or other animals" based on the Medical Device Amendments of 1976 (Hackett and Gutman, 2005). To determine the "intended use" that is so key to its regulation, the FDA considers a device maker's advertising, product distribution, labeling claims, product websites, and any form of promotional material on the product (Heller, 2006).

When the FDA asserts jurisdiction, this typically results in premarket submissions to the FDA under its premarket approval (PMA) or premarket notification (510[k]) requirements (Box 3-1). To help determine which route is most appropriate, the FDA evaluates how much risk the diagnostic poses, how it differs from other currently available diagnostics, and its intended use. Tests that pose the most risk, are the most innovative, or are intended "for a use which is of substantial importance in preventing

BOX 3-1 Premarket Approval and Premarket Notification at the FDA

A PMA application usually requires manufacturers to submit clinical data showing that their device is safe and effective for its intended uses. For some tests, these clinical data can be published clinical studies and/or practice standards that can help determine the clinical performance of the test or retrospective comparisons of the diagnostic's performance with that of another device that has already been clinically tested. But often the FDA requires prospective clinical studies to assess a new device's safety and effectiveness.

The Safe Medical Devices Act of 1990 authorizes the FDA to request data on clinical sensitivity, specificity, and predictive value for diagnostic tests that undergo a PMA review. These data are costly and time-consuming to procure, and they require clinical research expertise that many small companies lack. Most manufacturers try to avoid the necessity of a PMA review of their diagnostic tests and may even forgo bringing their test to market if a PMA application is required.

Manufacturers can bypass the need for a PMA application if they can show that their device is substantially equivalent to one already on the market. This qualifies their device to enter the market via a 510(k) review process. This review requires manufacturers to submit data showing the accuracy, reproducibility, and precision of their diagnostic. Manufacturers also have to provide documentation supporting their claim that the diagnostic is "substantially equivalent" to a device already on the market.

As is true for PMAs, there are no well-defined performance standards for 510(k) reviews, nor does the FDA clearly define the requirements for substantial equivalence. However, the agency has issued guidance documents that indicate the standards by which it will review a variety of types of diagnostics. It has also accepted the laboratory test standards set by other organizations, such as the Clinical Laboratory Standards Institute. None of these standards, nor the 510(k) or PMA review process itself, considers the clinical safety and effectiveness of the diagnostic.

SOURCES: Gutman, 2000; Hackett and Gutman, 2005; IOM, 2005; FDA, 2006a.

impairment of human health" are subject to the most regulatory scrutiny (Gutman, 2000; IOM, 2005).

Since the Medical Devices Act was enacted in 1976, a number of novel diagnostic tests based on genetics and other innovative molecular biology technologies have emerged. This created a large category of tests that would have had to undergo PMA review because there were no similar devices on the market on which to base a less onerous 510(k) review. The FDA Modernization Act in 1997 created a "de novo classification" for a device that is not equivalent to a legally marketed device. This classification allows manufacturers to bypass a PMA review for novel, low-risk devices. Such devices are reviewed for safety and efficacy by the FDA in a streamlined manner that usually does not require prospective clinical studies, relying instead on existing clinical literature to determine the device's safety and effectiveness (Hackett and Gutman, 2005).

A PMA or 510(k) review may not be required if a cancer biomarker test is developed by a laboratory for in-house use (a "homebrew" test). The FDA historically has not regulated homebrew tests, and laboratories offering them must label their test results with a qualifier that indicates the tests have not been cleared or approved by the FDA (FDA, 2003b). The homebrew exemption can enable manufacturers to quickly bring their tests to market. For example, there are hundreds of genetic tests currently on the market, but only four have been granted FDA approval (Hudson and Javitt, 2006).

In 1992, the FDA attempted to exert more regulatory control over homebrew tests via its compliance policy guideline, which proposed applying general medical device regulation to homebrew tests. But due to strong objections from the laboratory community, which claimed that the proposed guideline would be an onerous duplication of regulations promulgated under the Clinical Laboratory Improvement Amendments (CLIA, see next section), the FDA stated that "the use of in-house developed tests contributed to enhanced standards of medical care in many circumstances, and that significant regulatory changes in this area could have negative effects on the public health" (DHHS, 2003). However, the FDA asserted that it had the authority to regulate homebrew tests should it wish to do so (Shapiro and Prebula, 2003), and it has been suggested that the agency's choice not to regulate in-house tests was due to resource constraints (Heller, 2006).

Instead, the agency tried to ensure the safety and effectiveness of homebrew tests by regulating the building blocks, known as analyte-specific

---

BOX 3-2   FDA Regulation of Analyte-Specific Reagents

In a ruling made effective in November 1998, the FDA subjected both the manufacturers of ASRs, as well as the laboratories using them, to regulation to ensure that ASRs would be made consistently over time according to the agency's quality control requirements. It is the responsibility of the laboratory using the ASRs to develop a recipe for the homebrew test that incorporates the reagents, and it cannot share that recipe with other labs. All the testing using a homebrew diagnostic is done within the laboratory of the company or organization that developed it. Laboratories that produce ASRs must register with the FDA and satisfy the agency's Quality System Regulations (good manufacturing practices), as well as report postmarket device failures. They are also required to indicate on the label for the ASR that its analytical and performance characteristics are not established.

Makers of homebrew tests are not permitted to market their tests to other laboratories, nor can they sell packages of ASRs, or an ASR linked to a solid surface, with instructions on how to use the reagents in a testing procedure. Such packaging is considered to be a test kit subject to FDA review.

SOURCES: FDA, 2003b; Shapiro and Prebula, 2003; IOM, 2005.

---

reagents (ASRs),[1] for these tests (Hackett and Gutman, 2005) (Box 3-2). In response to requests from manufacturers to clarify ASR regulations, the FDA recently issued draft guidance to better explain how FDA defines ASRs and to more clearly delineate the regulatory rules of these products for ASR manufacturers (FDA, 2006c). The document also provides examples of entities that FDA does and does not consider to be ASRs.

In addition to guidance documents, the FDA has also used warning letters to assert authority and establish precedent for oversight of new tests that manufacturers thought would be outside FDA's jurisdiction. For example,

---

[1]These reagents are defined as "antibodies, both polyclonal and monoclonal, specific receptor proteins, ligands, nucleic acid sequences, and similar reagents, which through specific binding or chemical reaction with substances in a specimen, are intended for use in a diagnostic application for identification and quantification of an individual chemical substance or ligand in biological specimens" (FDA, 2003b).

the FDA recently prevented Roche Molecular Diagnostics from registering their new microarray genetic test for drug metabolism (AmpliChip CYP450) as an ASR. The denial was based, in part, on an assessment that the intended use of the AmpliChip to identify genetic indicators of drug metabolism capabilities "is of substantial importance in preventing impairment of human health" (FDA, 2003a). Furthermore, the FDA does not regard a microarray, which uses multiple reagents to detect a genetic profile, as falling under its definition of an ASR (Hackett and Gutman, 2005). The FDA suggested seeking de novo classification for their AmpliChip, and Roche's submission of previously published clinical literature on the genetic variants the AmpliChip detects and their clinically significant effects on drug metabolism led to FDA's approval of the AmpliChip (Hackett and Gutman, 2005).

The FDA has also required manufacturers of preanalytical systems, which collect, stabilize, and purify RNA, to submit a 510(k) premarket notification for the devices (FDA, 2005c), and it recently asked the makers of a new serum protein test using mass spectroscopy for ovarian cancer screening (OvaCheck) to consult with the agency about the appropriate regulatory status of the test. The developers of the OvaCheck test expected it would fall under the homebrew exemption from FDA review. But the FDA indicated that the test may be subject to a 510(k) review because the software used to analyze the results could be considered a device intended for use in the diagnosis of disease and therefore subject to regulation (FDA, 2004b).

In September 2006, the FDA issued draft guidance for such tests that use complex mathematical formulas to interpret large sets of gene or protein data, referred to by the FDA as In Vitro Diagnostic Multivariate Index Assays (IVDMIAs) (FDA News, 2006a). The document notes that "the manufacture of an IVDMIA involves steps that are not synonymous with the use of ASRs and that are not within the ordinary 'expertise and ability' of laboratories that FDA referred to when it issued the ASR rule. Therefore, IVDMIAs do not fall within the scope of laboratory developed tests over which FDA has generally exercised enforcement discretion."

The FDA also recently warned Access Genetics that several of the genetic test packages it manufactures and sells contain all that is needed to perform the tests, including lab assay protocols, and therefore are not homebrew tests, which are conducted only at the site at which they were developed (FDA, 2005e). The agency also notified the Nanogen Corporation that its NanoChip Molecular Biology Workstation, NanoChip Electronic

Microarray, and several ASRs were neither approved as a single system nor as separate components (FDA, 2005d; Heller, 2006). Furthermore, the FDA pointed out that some of the manufacturer's publicity about its NanoChip system indicated that it could be used for clinical diagnostic applications and therefore could not be considered a research-only diagnostic exempt from FDA review, as the company expected (FDA, 2005d). If a test is used for research only, the FDA does not exert jurisdiction, but if the assay is used for a clinical purpose, such as for diagnosis, it is subject to regulation by the FDA. Neither the FDA nor CLIA offers any clear guidelines, however, for distinguishing the difference between a research-only diagnostic and a clinical diagnostic (Hackett and Gutman, 2005; Heller, 2006).

The FDA also appears to be more assertive now in requesting clinical data for its reviews of biomarkers linked to therapeutics. Biomarkers used in clinical trials to identify likely responders to drugs (pharmacogenomic tests) will be regulated as devices in parallel with their corresponding drug candidates, and those for higher risk conditions will require PMAs. The FDA guidance (2005a) recommends submitting pharmacogenomic data when the data will be used to make approval-related decisions and when the data are relied on to define, for example, trial inclusion or exclusion criteria, the assessment for prognosis, dosing, or labeling or used to support the safety and efficacy of a drug. If a test shows promise for enhancing dosing, safety, or effectiveness or will be specifically referenced on the label, the FDA recommends codevelopment of the device and drug and coordinated applications for FDA approval (FDA, 2005a). The experience with attempts to add pharmocogenetic tests for the drug metabolizing enzyme cytochrome p450 to the labels for drugs such as warfarin indicate just how great the challenge of validation can be (IOM, 2006).

In addition, in its February 2006 draft guidance on pharmacogenetic tests and genetic tests for heritable markers, the FDA stated that "For predictive screening in healthy or asymptomatic individuals, long-term follow-up (i.e., a longitudinal study) may be the only way to prove that the test was indeed predictive and to evaluate issues such as penetrance" (FDA, 2005a, p. 4). But this guidance also noted that for some genetic tests, there may be a sufficient clinical literature base to establish clinical validity of the new test without extensive new clinical studies.

CMS OVERSIGHT OF
CLINICAL LABORATORY PERFORMANCE

Laboratory performance is overseen by the Centers for Medicare & Medicaid Services (CMS)[2] under the Clinical Laboratory Improvement Amendments of 1988 (Hackett and Gutman, 2005). To be operational, a laboratory that conducts testing on human specimens for the purpose of providing information relevant to the diagnosis, prevention, or treatment of disease or physical impairments or for health assessments must be CLIA-certified (FDA, 2006b; Javitt, 2006). CLIA certification, which is renewed contingent upon inspection every two years, is intended to ensure the accuracy, reliability, and timeliness of patient test results from laboratories throughout the United States (FDA, 2006b; Box 3-3). But CLIA does not replace FDA regulatory authority over medical diagnostic tests; it does not address the clinical accuracy or usefulness of tests.

There are some state requirements that are more stringent than CLIA, as well as organizational guidelines and standards that can be voluntarily adopted by laboratories to further the accuracy of their testing (DHHS, 1999; Swanson, 2002). But most laboratories follow the minimum generic standards set by CMS under CLIA. The requirements for CLIA certification vary depending on whether laboratories conduct tests of moderate or high complexity. (Low-complexity tests, such as a urine dipstick, are simple enough to be performed by unskilled laboratory personnel or even by patients. These tests are waived from requiring CLIA certification.) The FDA determines the degree of complexity of in vitro diagnostics based on the amount of expertise, oversight, interpretation, and judgment required to perform the test, as well as the potential risk to public health if the test is inaccurately performed (FDA, 2006b).

---

[2]The Centers for Disease Control and Prevention (CDC) provides scientific and technical support to CMS and convenes the Clinical Laboratory Improvement Advisory Committee. The Committee provides scientific and technical advice and guidance regarding the need for, and the nature of, revisions to the standards under which clinical laboratories are regulated; the impact on medical and laboratory practice of proposed revisions to the standards; and the modification of the standards to accommodate technological advances. The Committee consists of 20 members knowledgeable in the fields of microbiology, immunology, chemistry, hematology, pathology, and representatives of medical technology, public health, clinical practice, and consumers. In addition, the Committee includes three ex officio members: the Director, Centers for Disease Control and Prevention; the FDA Commissioner; and the CMS Administrator.

BOX 3-3   Overview of CLIA Regulation of High- and Moderate-Complexity Tests

I.   **Demonstration of performance specifications**

 a. Assess day-to-day, run-to-run, and within-run variation
 b. Verify test-reporting ranges from published reference ranges, kit manufacturer's ranges, or in-house testing
 c. Provide evidence of reproducibility

II. **Quality control**

 a. Create a procedure manual for each test
 b. Perform calibration procedures at least once every 6 months for each test system
 c. Perform quality control each day a test system is used, with at least two levels of control (high and low limits)

III. **Proficiency testing (PT)**

 a. A lab must enroll at least one person in a PT training program for each specialty for which it seeks certification; ideally, lab members will rotate for each PT specialty or subspecialty testing program
 b. PT samples must be tested in the same manner by the same personnel as patient samples

---

Both moderate- and high-complexity tests require laboratories to document the accuracy and reproducibility of their testing, the use of quality control procedures, and the proficiency training and testing of key personnel (Swanson, 2002; DHHS, 2003; CMS, 2006). The main difference between high- and moderate-complexity laboratories is that there are more stringent qualifications required for the personnel of high-complexity laboratories (DHHS, 2003). The FDA's ASR ruling restricts sale of a reagent used for clinical purposes to laboratories designated as high complexity under CLIA, because these labs were thought to have the personnel and systems in place to allow for the reliable development of in-house tests.

Most moderate- to high-complexity tests fall under CLIA-specified specialty areas that require more specific proficiency testing programs. These include tests within the domains of microbiology and immunology.

c. The PT should provide a minimum of 5 samples per testing event with at least 3 testing events per year

d. Level of accuracy for satisfactory performance for PT testing varies depending on the analyte(s) involved

IV. **Personnel**

a. Labs must identify qualified individuals for the following positions:

i. *Moderate complexity:* director, technical consultant, clinical consultant, and testing personnel

ii. *High complexity:* director, technical supervisor, clinical consultant, general supervisor, testing personnel

b. Personnel qualifications differ depending on the position and level of test complexity; an MD, DO, DPM, or PhD with the appropriate laboratory training and experience can fill all the required positions for both complexity levels

c. Labs must keep credentials of every lab member on file for inspection

Sanctions

a. Sanctions include suspension, limitation, or revocation of CLIA certificate, Medicare payment approval cancellation, civil money penalties, onsite monitoring, and correction plan

SOURCE: CMS, 2005.

But there are no specialty areas requiring proficiency testing indicated for molecular or biochemical genetic testing, despite the recognition by Congress that proficiency testing "should be the central element in determining a laboratory's competence, since it purports to measure actual test outcomes rather than merely gauging the potential for accurate outcomes" (DHHS, 1988; CMS and DHHS, 2004; Javitt, 2006).

In 2000, the Secretary's Advisory Committee on Genetic Testing generated a report that concluded that the oversight of genetics tests was insufficient to ensure their safety, accuracy, and clinical validity and recommended that CMS develop a specialty area for genetic testing under CLIA. Draft guidelines for genetic testing quality from the international Organisation for Economic Co-operation and Development also identified proficiency testing and lab quality as critical to ensuring health (OECD,

2006). In March 2006, the Genetics and Public Policy Center of Johns Hopkins University conducted a survey of laboratory directors of genetic testing laboratories and found widespread support for creation of a genetic testing specialty under CLIA (Hudson et al., 2006). Furthermore, the survey data found that proficiency testing was linked to greater accuracy of genetic testing, although at least a third of genetic testing laboratories fail to perform proficiency assessments for some or all of their tests. In April 2006, CMS proposed rule making that would create such a specialty area, but three months later CMS stated that existing CLIA regulations are adequate to protect public health, asserting that there is insufficient "criticality" to warrant rule making for genetic testing (Genetics & Public Policy Center, 2006; Hudson, 2006). The committee agrees with the need for oversight and recommends developing a specialty area for molecular diagnostics.

THE NEED FOR CONSISTENCY AND TRANSPARENCY

Clearly there is significant variability in the scrutiny of biomarker tests before and after entry into the market. This lack of consistency and transparency in the biomarker development process is problematic for two important reasons. First, the variability and uncertainty associated with oversight and assessment of biomarker tests are disincentives to innovation by developers. As noted above, the FDA previously has claimed legal authority to assert jurisdiction over diagnostic tests, but it has usually withheld its authority. Recently, the FDA has taken action to create clarification and precedent on a case-by-case basis regarding molecular diagnostics through letters or guidance documents. But when oversight is variable, evolving, and thus hard to predict, it can have a major impact on the risk of development. Unanticipated action by the FDA can result in delays and greatly increase the cost of development. As noted in Chapter 4, the variability and unpredictability of health care coverage adds an additional layer of risk and uncertainty for developers. Once a test enters the market, coverage decisions often depend on convincing evidence of clinical usefulness, but those decisions are made on an ad hoc basis and vary by payor, as there are no widely accepted guidelines for evidence standards.

Second, a lack of regulation and consistent assessment prior to market entry can lead to inappropriate adoption and use of biomarkers, unnecessarily increasing health care costs and potentially harming patients. Many diagnostic tests in use have not been validated or formally evaluated. Companies develop their own assessment criteria and standards for developing and marketing diagnostic tests on a case-by-case basis and generally choose the path to market of least resistance. Competition tends to erode standards of evaluation, since the more rigorous the standard, the longer and more costly the development process and the less likely it is to be first to market. Most diagnostic tests enter the clinic as homebrew tests, which are exempt from FDA approval or clearance. Furthermore, even if a company seeks and obtains FDA approval, laboratories can develop their own in-house homebrew test and use that in place of the FDA-approved test. No federal agency currently enforces the accuracy of marketing claims made for homebrew tests.

Thus, there is a great need for a coherent strategy to make the biomarker development and adoption process more transparent, to remove inconsistency and uncertainty, and to elevate the standards and oversight applied to biomarker tests. No federal agency currently takes responsibility for ensuring the clinical validity of biomarkers, but oversight and ownership of the process are key to developing strategies and making effective and efficient progress in the field. The committee strongly urges the designation of an appropriate federal agency to provide leadership in the process and to coordinate and oversee interagency activities. The National Institute of Standards and Technology (NIST) is an appealing candidate. Although it has had a limited role in biomarker development to date due in part to financial restraints, it has the appropriate experience to play a broader role in the establishment of standards for biomarkers if given appropriate funding for that purpose. NIST standards work in health care and clinical chemistry is well established, and, more recently, NIST has begun some work related to cancer molecular genetics technology and standards, as well as work with the Early Detection Research Network of the National Cancer Institute (NCI) on cancer biomarker validation (Barker, 2003).

An important first step would be to convene all relevant government agencies (e.g., the National Institutes of Health [NIH], the FDA, CMS, the Agency for Healthcare Research and Quality, NIST) and non-government stakeholders (e.g., academia, the pharmaceutical and the diagnostics industry, and health care payors) to work together in developing a transparent process to create well-defined consensus standards and guidelines for biomarker development, validation, qualification, and use to reduce the variability and uncertainty in the process of development and adoption. For example, FDA, CMS, and industry should work together to develop guidelines for clinical study design that will enable sponsors to run a single study (or a minimal number of studies) to generate adequate clinical data for review by both agencies. Optimizing clinical study design in this way could shorten time to market, reduce cost and risk, and strengthen the evidence base for evaluation. The FDA noted the importance of standards when formulating its Critical Path Opportunities List in 2006, by including several projects aimed at devising standards for microarray and proteomics-based identification of biomarkers, mapping the process and criteria for qualifying biomarkers for use in product development, and developing clinical trial data standards (DHHS and FDA, 2006).

Developing a complete set of guidelines and standards is an ambitious goal, as different guidelines will probably need to be developed for different stages of the development pathway, for different applications, such as for the different stages of drug development and clinical application (e.g., screening, diagnosis, treatment planning, response monitoring, and surrogate endpoints), for different technologies, and for single biomarkers versus panels or patterns. A flexible and adaptable process for monitoring the guidelines will also be needed so that they can be revised as technologies or evidence change, and they will probably require regular review and updates.

There are many informative examples that could serve as precedents to guide this process, as many professional organizations and collaborative groups have already proposed guidelines for various steps in the process (Box 3-4). These initiatives provide an excellent starting point, but there are numerous gaps in the continuum of biomarker development, adoption, and use that need to be filled. In addition, with so many sources of standards development, there is potential for overlap, with competing or conflicting standards that could lead to confusion. The situation could be greatly improved if a single entity took responsibility for providing overarching leadership in the area of biomarker development and use. Furthermore, adherence to most of these guidelines is voluntary, so it is important to devise strategies to ensure compliance, including both incentives and penalties. For example, reporting standards of the Standards for Reporting of Diagnostic Accuracy or STARD initiative (NCI Division of Cancer Prevention, 2006) have been adopted by a range of journals in reporting results of clinical studies of diagnostic tests. Similarly, the Minimum Information About a Microarray Experiment (MIAME) guide (MGED Society, 2005), developed by the Microarray Gene Expression Data Society, defines requirements for effective reporting on the entire process of collecting, managing, and analyzing microarray data so that the data can be reused and interpreted by others. The MIAME guide has been adopted by several scientific journals as a requirement for publication, although the stringency of compliance enforcement by the journals likely varies. Broad adherence to these guidelines would be ensured if required for receipt of funding from federal agencies, including NIH.

It is also critically important for the FDA to clarify its authority over biomarker tests linked to clinical decision making and then establish and consistently apply clear guidelines for compliance with the requirements of biomarker test oversight. The committee recognizes that the FDA has limited resources and that additional funding will be necessary to make meaningful changes in this arena. That said, it should be noted that expanding FDA regulation to all homebrew tests would not be a wise use of limited FDA resources, and it is not desirable. However, it is desirable for the FDA to define a set of criteria for molecular diagnostic tests that would trigger additional oversight for those tests that are complex and are most likely to have an impact on the public's health. The process for establishing these criteria and the associated regulations will need to be dynamic in order to adapt to rapid changes in technology.

The committee also strongly recommends that the appropriate federal agency (the Federal Trade Commission [FTC] or the FDA) effectively monitor and enforce marketing claims made for molecular diagnostics. The FDA's ASR ruling limits ordering of homebrew tests using ASRs to a health professional or "other persons authorized by state law." But the regulation does not preclude a health professional who is an employee of the company that offers a homebrew test from ordering the test for patients (Hudson and Javitt, 2006). CLIA oversight also does not restrict when and for whom a test may be performed, unlike the labels for FDA-approved drugs, which must specify indications for use in order to enter the market (Hudson and Javitt, 2006). This lack of regulatory definition of who may authorize the use of a homebrew test and for whom has opened the door for direct-to-consumer advertising of homebrew tests and their use by consumers without consulting their physicians.

At least eight companies currently promote genetic testing for health-related conditions directly to consumers through their websites and more are expected to in the future (Hudson and Javitt, 2006). The accuracy of this direct-to-consumer advertising of homebrew tests is not being regulated by the FDA, which has regulatory authority over advertising claims only for the products it reviews and approves or clears. The FTC prohibits false or misleading advertising and has claimed it will take action against such advertising of genetic tests. But the agency's limited resources appear to be

BOX 3-4   Examples of Standards and Guidelines for the Development and Use of Biomarkers

Microarray Gene Expression Data Society

The initial goal of the Microarray Gene Expression Data (MGED) Society was to create standards for presenting and exchanging microarray data in order to improve the quality and reproducibility of microarray studies. MGED was founded as a grassroots movement in November 1999 and transitioned into an international nonprofit organization in 2002. MGED seeks to establish standards for microarray data annotation and exchange, facilitate the creation of microarray databases and related software, and promote sharing of microarray data. MGED plans to expand these goals to other functional genomics and proteomics high-throughput technologies in the future. Member organizations meet to exchange information and discuss goals in annual international conferences. MGED also supports seminars and tutorials for programmers and others interested in microarray design, quality control, and more. Seminars and conferences encourage attendants to contribute suggestions and improvements for MGED projects.

MGED is currently pursuing six major projects in the form of working groups that conference via closed e-mail discussion boards. One major product of MGED is the creation of guidelines for the Minimum Information About a Microarray Experiment that should be reported so that others may unambiguously repeat and interpret microarray experiments. MIAME's focus is on the content and structure of microarray information rather than on the format for capturing that information. While it serves as a guide to the development of microarray databases and data management software, it does not address different types of experiments. MIAME recommends that all reported microarray experiments provide annotation of samples, reliability estimates for particular data points, and standardized vocabularies and ontologies. MIAME is now a requirement for publication in several scientific journals, including the *Nature Group*, *The Lancet*, *Cell*, and *EMBO Journal*. The guidelines have also inspired the development of standards in other fields, including metabolomics and proteomics.

SOURCES: Brazma et al., 2001; IOM, 2006; MGED, 2006.

Microarray Quality Control Project

Sparked by a 2003 paper by Margaret Cam that showed significant inconsistencies in microarray data across different platforms, the FDA initiated the Microarray Quality Control (MAQC) Project in 2005. The MAQC is part of the FDA's Critical Path Initiative to modernize the scientific process by which potential drugs, medical devices, and biological products are developed into medical products. The MAQC Project seeks to validate microarray technology and publish standards for data from microarrays and other technologies, such as QRT-PCR, that will be made available to the microarray community. These standards will define thresholds and quality measures that can be used to assess the precision and comparability of data across different technology platforms. Such comparisons should help to identify and eliminate any systematic biases that may exist between microarrays and QRT-PCR. The MAQC Project is unique in that it is larger and more comprehensive than other comparisons and datasets generated thus far. The project involved six FDA centers, major producers of microarray platforms and RNA samples, the Environmental Protection Agency, the National Institute of Standards and Technology, and academic research centers. A total of 20 microarray products and 3 alternative technologies were used to perform over 1,300 tests at different labs. A total of 1,329 microarrays were used in the project (for a complete list of all microarrays used in the project, go to *http://www.fda.gov/nctr/science/centers/toxicoinformatics/maqc/docs/ MAQC_Summary_1stPhase.pdf*).

The results of the MAQC Project show that, overall, levels of variation between microarray runs at different sites and with different platforms were relatively low (total coefficient of variance ranged from 10 to 20 percent) and reproducibility was high (expression results overlapped 70–90 percent of the time). Although these results suggest that microarray technology may be more reliable for clinical applications than previously thought, some caution that the MAQC studies were conducted in "optimized" settings that may be difficult to recreate due to time and sample processing restrictions. Moreover, some argue that the vast majority of variability in the data is biological rather than technical. However, researchers can now compare their microarray technologies against the MAQC data to assess their genomic data quality.

MAQC Project members met in September 2006 to discuss how to make microarray data useful in clinical settings. The MAQC results were released to the public on September 8, 2006, and published in the September 2006 issue of *Nature Biotechnology*. Those articles, which summarized MAQC findings and data sets, can be viewed free of charge at *http://www.nature.com/nbt/focus/maqc/index.html*. The FDA plans to publish guidance on microarray quality control and data analysis in December 2007.

SOURCES: Tan et al., 2003; Couzin, 2006; DHHS, 2006; FDA, 2006d; Frueh, 2006; Perkel, 2006a,b.

continued

BOX 3-4   Continued

Human Proteome Organization Proteomics Standards Initiative

The Human Proteome Organization (HUPO) Proteomics Standards Initiative (PSI), founded in 2002, aims to define community standards for data representation in proteomics to facilitate data comparison, exchange, and verification. It has numerous working groups that consist of academic, government, and industry researchers, software developers, publishers, and instrument manufacturers. The groups are developing a set of Minimum Information About a Proteomics Experiment (MIAPE) documents to provide guidelines on how to adequately report on various types of proteomics experiments. HUPO plans to eventually publish these documents, with the expectation that the requirements within will be enforced by journals, compliant repositories, and funders. The group is also working to develop formats for data exchange, as well as standardized vocabularies and ontologies.

SOURCE: HUPO, 2006.

American Society for Biochemistry and Molecular Biology Criteria for Publication of Proteomics Data

The American Society for Biochemistry and Molecular Biology held a workshop in May 2005 to create a standardized set of criteria for the publication of proteomic data so that the entire proteomics community, including both specialists and nonspecialists, could confidently understand and use the standards for acceptable proteomics data.

Attendees included members of the editorial advisory boards of major publishing groups and journals focusing on proteomics. Participants were divided into subgroups that were assigned different aspects of the guidelines. The guidelines created by editors of the *Molecular and Cellular Proteomics* journal, published in 2004, were used as a framework on which to build. During the workshop, each subgroup drafted and presented a preliminary set of criteria.

The final criteria were published in March 2006. Proteomics guidelines for publication are now posted on the websites of *Proteomics*, *Molecular and Cellular Proteomics*, and the *Journal of Proteome Research*. The enforcement policies for adherence to the criteria are at the discretion of the journal's editorial staff.

SOURCES: Beavis, 2005; Cottingham, 2005.

NIH Workshop: Standards in Proteomics
Bethesda MD, January 4–5, 2005

The Standards in Proteomics workshop was a component of the Building Blocks, Biological Pathways, and Networks subdivision of NIH's Roadmap for Medical Research initiative. The general goal of the meeting was to develop a community-based plan for consistent analysis, representation, dissemination, and publication of proteomic data. Participants also discussed a variety of logistical and technical issues related to implementing standardized data repositories for proteomics experiments. No definitive guidelines were derived from the meeting, but specific goals included the following:

- Create a plan that will guide the scientific community—including researchers, funding agencies, and journals—toward establishing standards for the description of proteomic experiments and data presentation.
- Establish mechanisms and strategies for implementing standards.
- Discuss the status of available software tools for proteomic data analysis and draft strategies for dissemination and support of open-source proteomic software tools.
- Create a plan for implementing guidelines for publication and presentation of proteomic data and experiments.
- Generate a summary document that describes guidelines and available tools for proteomic data publication and presentation.

SOURCE: NIH, 2006.

Standards for Reporting of Diagnostic Accuracy

The Standards for Reporting of Diagnostic Accuracy (STARD) evolved out of a 1999 Cochrane Colloquium meeting at which the Cochrane Diagnostic and Screening Test Methods Working Group noted the substandard reporting of diagnostic test evaluations. To improve the accuracy and completeness of reporting of studies on diagnostic accuracy, the working group created a steering committee of international experts to assess what should be included in a study report on diagnostic accuracy that would allow the reader to detect potential bias in the study, as well as to assess the applicability of the results. The steering committee did an extensive search of publications on the conduct and reporting of diagnostic studies and then convened a consensus meeting that included researchers, editors, methodologists, and professional organizations.

continued

BOX 3-4   Continued

The result was a 25-item general checklist and a flow chart that together help authors describe the essential elements of their design and conduct of the study, the execution of tests, and the results. The checklist specifies exactly what is needed for the title, abstract, introduction, methods, results, and discussion sections of a journal article written about a diagnostic study. The flow diagram indicates visually the process of sampling and selecting participants, the portion of participants that received the test or reference standard, and the portion of patients at each stage of the study.

The STARD group thought that their general checklist for reporting studies of diagnostic accuracy applicable to research in any field would be more likely to be adopted by authors, peer reviewers, and journal editors than different checklists for each field. STARD was published in 2003 and offers voluntary guidelines for researchers and reviewers of research articles related to diagnostics tested clinically, although some journals have already adopted them as requirements for publication.

SOURCE: Bossuyt et al., 2003.

NIST–EDRN Workshop on Standards and Metrology for Cancer Diagnostics

The National Institute of Standards and Technology and the Early Detection Research Network (EDRN) jointly sponsored a workshop in August 2005, with the aim of comparing the performance characteristics of different analytical platforms; to assess the needs for standard methods, assays, and reagents for cancer biomarker development and validation; and to make recommendations for the development of standard reference materials and standard operating procedures. The workshop focused on the following areas:

- Methods and standards to quantitatively and reliably measure DNA methylation in clinical specimens.
- Standard proteomic reference materials for cancer biomarker discovery and validation and for cross-validation between laboratories and platforms.
- Characteristics of clinical reference materials that can be used to accelerate the discovery and validation of cancer biomarkers.

This workshop was one component of a broader effort on the part of EDRN to develop and test standards and paradigms for early cancer detection biomarkers.

SOURCES: NCI Division of Cancer Prevention, 2005, 2006; Barker et al., 2006.

Receptor and Biomarker Group of the European Organization for Research and Treatment of Cancer

The Receptor and Biomarker Group (RBG) of the European Organization for Research and Treatment of Cancer (EORTC) is comprised of cancer researchers and clinicians from 18 European countries and primarily serves the European Community. The RBG-EORTC tries to establish and maintain the analytic and clinical validity of tumor biomarker tests by establishing quality assurance schemes that are obligatory for all markers used in the EORTC clinical trials, as well as by having its experts advise clinical cancer researchers on appropriate methodology and interpretation of results for tumor assays. The RGB-EORTC provides guidelines for the assay performance and handling of test materials collected in retrospective or prospective clinical trials and also provides procedures for preclinical laboratory testing. The group also provides biomarker laboratory reference materials to aid standardization of biomarker assays, as well as its own sensitive and specific assays for a number of biomarkers, including urokinase-type plasminogen activator and vascular endothelial growth factor. On a regular basis, the RBG-EORTC evaluates new tumor biomarkers and makes recommendations to international certifying boards and also to the European Commission for Registration of Biological Reagents. In addition, the RBB-EORTC tests new commercial kits for existing biomarkers and evaluates them, when appropriate, against currently accepted assays.

SOURCE: Schmitt et al., 2004.

NCI-EORTC REMARK (Reporting of Tumor MARKer Studies) GUIDELINES

Recognizing that the number of biomarkers that become clinically useful is "pitifully small" compared with the number of reports on tumor markers, these 2005 guidelines were developed by the Statistics Subcommittee of the National Cancer Institute-European Organization for Research and Treatment of Cancer (NCI-EORTC) working group. The goal of the guidelines is to improve the reporting standards for published clinical tumor marker studies to allow adequate assessment of the quality of the study and the generalizability of study results and to improve the ability to compare results across studies.

These are voluntary guidelines for researchers and reviewers of journal articles, although some journals may require adherence to the guidelines for publication. The guidelines focus on what should be reported for studies on clinical prognostic markers (those that predict

continued

BOX 3-4 Continued

clinical outcomes irrespective of treatment), although some of its require-ments are also relevant to studies on predictive markers (those that predict response to specific treatments) and to tumor markers that are early in their development and have yet to be applied in a clinical setting. The REMARK guidelines were developed mainly for studies that evalu-ate single tumor markers and are not applicable to genomic or proteomic studies that simultaneously evaluate large numbers of markers.

To develop the guidelines, the NCI-EORTC working group, comprised of statisticians, clinicians, and laboratory scientists, considered literature citing inadequate reporting or problematic analysis methods in published studies of tumor markers, as well as similar reporting guidelines devel-oped for other types of medical research studies. The guidelines do not have specifications unique to tumor markers or the technologies used in their assays, but rather they list the relevant information that researchers should provide about their study objectives, materials and methods, study designs, statistical analyses, and results. The guidelines also suggest helpful presentations and analyses of data and require that a discussion include the limitations of the study and the clinical value of its results. As the working group noted, "high-quality reporting of a study cannot transform a poorly designed or analyzed study into a good one, but it can help to iden-tify the poor studies, and we believe it is an important first step in improving the overall quality of tumor marker prognostic studies" (p. 9068).

SOURCE: McShane et al., 2005.

American Society of Clinical Oncology Clinical Practice Guidelines for the Use of Tumor Markers in Breast and Colorectal Cancer (2000 updated version)

The American Society of Clinical Oncology (ASCO) developed con-servative voluntary clinical guidelines for physicians that recognize their ultimate use/application depends on the physician's judgment, taking into consideration each patient's individual circumstances.

To determine the clinical suitability of a cancer biomarker, the guide-lines committee used the medical literature to evaluate how six tumor markers for colorectal cancer and eight for breast cancer affected such clinical outcomes as overall survival, disease-free survival, quality of life, toxicity, and cost-effectiveness of treatment. The guidelines committee considered the strength of the evidence from each study based on the quality of the study, with the most weight placed on evidence gathered from large, prospective, randomized controlled clinical trials. All clinical uses of the biomarker were considered, including screening, diagnosis, staging, surveillance, and monitoring response to treatment.

The guidelines committee disregarded strong correlations between disease progression or disease response and a specific result on a biomarker test if physicians could not reliably use the result to alter a clinical course. For example, blood levels of the biomarker CA 27.29 tend to increase as breast cancer progresses, so that one well-designed study found it could detect recurrence about 5 months, on average, before other symptoms or tests. But that ability did not change therapy options or show a documented affect on disease-free or overall survival. Consequently, the guidelines do not recommend using CA 27.29 as a monitoring tool for breast cancer recurrence. Of the biomarkers, the committee recommended only the clinical use of hormone receptor (ER, PR) or HER-2 status of breast tumors to determine treatment.

SOURCE: Bast et al., 2001.

Tumor Marker Utility Grading System (TMUGS) and TMUGS-Plus

The large number of preliminary studies, but few definitive studies, on tumor markers prompted some members of the ASCO committee that developed its practice guidelines for the use of tumor markers in breast or colorectal cancer to draft a clinical tumor marker utility grading system (TMUGS), which was published in 1996. The purpose of the TMUGS is to help research scientists design studies that will provide clinically useful data on tumor markers and to help expert reviewers evaluate published studies on tumor markers.

The TMUGS requires researchers to specify how the tumor marker is evaluated, provide materials and methods details, and detail how the marker can be used clinically. From this it gives a 0 to 3 rating of the relative utility of a tumor marker for a specific use and outcome. Only those that received 2+ or 3+ were included in the ASCO guidelines. To get these grades, tumor markers had to be reliable and provide information that clinicians could use in their decision making. The reliability of the tumor marker was based on the quality of the studies assessing it, with more reliability attached to tumor markers evaluated in large, prospective, randomized studies or in meta-analyses of studies that provide lower levels of evidence.

The TMUGS-Plus system was developed by British and American researchers and published in 1998. This system builds on TMUGS with the addition of a decision matrix in which weak, moderate, and strong prognostic categories are intertwined with weak, moderate, and strong predictive categories to enable reviewers to consider both when determining the clinical utility of a given tumor marker. The relative strength

continued

BOX 3-4　Continued

category in which a prognostic or predictive marker is placed is determined by how much it moves patients between prognostic stages or treatment response outcomes, respectively.

TMUGS was designed not only to aid expert assessments of published data regarding the clinical utility of tumor markers, but also to help clinical investigators design tumor marker studies that will reveal the clinical utility of the marker. The authors state: "We do not suggest that this system is useful for application of a factor to an individual patient's situation. Rather, we propose that the TMUGS-Plus system can be used to determine whether available data support the introduction of a tumor marker into routine clinical use. The individual physician and patient will then need to decide if the marker data are relevant to her particular situation" (p. 408).

Both TMUGS and TMUGS-Plus are voluntary systems for evaluating clinical studies of tumor markers.

SOURCE: Hayes et al., 1996, 1998.

College of American Pathologists Conference on Solid Tumor Prognostic Factors

The College of American Pathologists (CAP) convened a conference in 1999 to examine prognostic and predictive factors in breast, colon, and prostate cancers, with the aim of stratifying these factors into categories based on the strength of published evidence. Conference goals included reducing variation in methods, interpretation, and reporting, as well as developing strategies for implementing changes in how prognostic and predictive factors are evaluated and used. Working groups focused on cancer type–specific issues as well as issues common to all solid tumors.

SOURCE: Hammond et al., 2000.

Evaluation of Genomic Application in Practice and Prevention

More than 1,200 genetic tests for diseases have been developed, and 950 are available for clinical use. Concerns over the safety and utility of these tests prompted the initiation of the Evaluation of Genomic Application in Practice and Prevention (EGAPP) in fall 2004 by the Office of Genomics and Disease Prevention at the Centers for Disease Control and Prevention (CDC). EGAPP draws from prior work conducted at the CDC by the ACCE projects (the name is derived from the four components of evaluation—analytic validity, clinical validity, clinical utility and associated ethical, legal and social implications), which proposed and tested a system for collecting, analyzing, and disseminating existing data on the safety and efficacy of DNA-based genetic tests.

The overarching goal of the project is to develop a coordinated process for evaluating genetic tests and other genomic applications that are in transition from research to clinical and public health practice. In April 2005, an interagency steering committee of the Department of Health and Human Services established a nonfederal, independent working group of 13 multidisciplinary experts. The EGAPP working group is charged with providing clear linkages between scientific evidence and subsequent recommendations for genetic tests by serving on technical review panels, providing guidance on projects, establishing methods and protocols, and selecting topics for review. Current topics under review include:

- Hereditary nonpolyposis colorectal cancer (HNPCC) screening
- Genomics tests for ovarian cancer detection and management
- Cytochrome P450 (CYP450) polymorphism testing in adults with depression
- UGT1A1 testing in colorectal cancer patients treated with irinotecan

The first three evidence reviews are being conducted by the Agency for Healthcare Research and Quality (AHRQ) Evidence-based Practice Centers, and the fourth is a more targeted review by a technical contractor.

At a recent meeting (June 2006), the working group presented subcommittee draft reports and decided on final topics for the third year. The final expected products from the working group are three to five major reviews, two to three fast-track reviews, and a document on methods and evaluation.

A long-term goal of the EGAPP is to create a sustainable process for pre- and postmarket assessment of genetic tests and other genomic applications in the United States. A critical component of EGAPP's work is making these reviews and subsequent recommendations accessible to the public. One of EGAPP's project activities is to develop informational messages that are targeted to specific audiences that would find the recommendations most relevant and useful. This information is intended to aid health care providers, payers, consumers, and policy makers to make informed decisions about the safety and efficacy of genetic tests and to safeguard against tests that may be released prematurely.

SOURCE: CDC, 2006.

preventing it from following through on its commitment. The FTC has yet to interfere with the direct-to-consumer claims made by the manufacturers of genetic tests, some of which appear to be false and misleading, according to some observers (Hudson and Javitt, 2006). Although the FTC, the FDA, and CDC recently issued a public alert to consumers about direct-to-consumer marketing of genetic tests, the message did not indicate that any actions were planned by any of those agencies (FTC, 2006).

Effective postmarket surveillance will also be needed to ensure the quality and accuracy of diagnostics, regardless of whether a biomarker test enters the market as a homebrew or with FDA approval or clearance. Although CLIA was intended to ensure the quality and accuracy of clinical laboratory tests, CLIA oversight appears insufficient to guarantee the accurate measurement and reporting of biomarker tests results. Experience with two well-established, prototypical cancer biomarkers that guide therapy

---

**BOX 3-5   Estrogen Receptor—
The Classic Cancer Biomarker**

For many decades the estrogen receptor (ER) has been used as a prognostic factor and as a predictor of response to endocrine therapies for breast cancer. It is considered a category I breast cancer prognostic factor by the College of American Pathologists, meaning that it is of proven prognostic importance and is useful in patient management. As such, accurate and reliable assessment of ER status is paramount for optimal breast cancer care. The usefulness of ER as a predictor of therapeutic response was first noted by retrospective review of patient data from many clinical trials conducted around the world. From those trials, 436 patients were identified in which the treatment response was recorded and ER was measured using one of several techniques, generally entailing some form of quantitative biochemical binding assay on fresh tumor samples. The results indicated that 55–60 percent of patients positive for ER responded to endocrine therapy, while those who were negative for ER had virtually no chance of responding.

Since the late 1990s, a semi-quantitative test based on immunohistochemistry (IHC) has become the method of choice, primarily because of its ease of use, reduced cost, and the ability to perform the assay on small samples of fixed tissue. Although studies have shown that IHC is equivalent or superior to binding assays (which are also known to produce false results), there is widespread concern that the variability

decisions for breast cancer patients—the estrogen receptor and the HER2 receptor—demonstrate the enormous challenges associated with the standardization and quality assurance of such tests (IOM, 2006; Boxes 3-5 and 3-6). There is a great deal of variation in the way these markers are measured and reported, and studies indicate a high rate of inaccurate results.

Surveillance and quality assurance activities, perhaps including proficiency testing, data collection and review, and/or inspections, could potentially be overseen by either the FDA or CMS and might be financed via user fees. A quality assurance testing program in place in the United Kingdom is an example of a possible model. This best practice program allows laboratories to compare their performance against reference materials and other laboratories and hence identify whether they have a testing problem (Ellis et al., 2004). The accompanying educational material and instructional assistance allows most laboratories to identify and rectify their

---

and inaccuracy of the test and interpretation of the results may lead to an unacceptably high error rate in determining ER status. The positive predictive value of ER tests is estimated to be in the range of 60 to 80 percent. The ER IHC test is not standardized, and many laboratories use FDA-approved reagents in different ways. In addition, there is no universal consensus on a scoring system for interpreting the results. A number of suggestions have been made to improve the reliability of the test results, including further improving and standardizing test kit reagents and controls, staining procedures, and scoring methods. Automated image analysis could perhaps also lead to more consistent and accurate results, but software programs must first be standardized and validated as well.

The situation is likely to become even more complex as newer methods have been developed to measure ER mRNA rather than protein. However, these methods are not yet widely used, in part because most labs are not equipped to conduct the tests, and they also are not fully standardized. Furthermore, new endocrine therapies, including aromatase inhibitors, are now available to treat breast cancer, and emerging evidence suggests that optimal response to a particular endocrine therapy depends on the level of ER expression, not just whether it is positive or negative.

SOURCES: McGuire, 1975; Fitzgibbons et al., 2000; Diaz and Sneige, 2005; Ross, 2005.

BOX 3-6   Herceptin/HercepTest
Development and Approval

The drug Herceptin (trastuzumab) targets the HER2 protein (human epidermal growth factor receptor 2), which is overexpressed in about 25 percent of breast cancer cases due to amplification of the gene. It has been widely noted that the efficacy of trastuzumab could not have been demonstrated in clinical trials in the absence of a selective biomarker to identify the patients most likely to respond (those with elevated expression of HER2). Mathematical models indicate that the clinical efficacy of the drug would have been difficult, if not impossible, to demonstrate with the number of patients typically recruited for a clinical trial if the study population had not been enriched with responders via a biomarker test for HER2. However, the assay used by Genentech during the clinical trials of trastuzumab was deemed inadequate for commercialization, so the company approached Dako Corporation to co-develop a commercial immunohistochemistry kit (HercepTest). The kit was then validated by demonstrating equivalence to the clinical trial assay. The FDA approved both Herceptin and HercepTest in September 1998. Dako then launched a comprehensive education program for pathologists. Because evidence shows that women with high levels of HER2 overexpression are more likely to respond to trastuzumab, accurate reporting of patient test results is crucial for therapeutic decision making.

However, studies have reported considerable variability in the accuracy of the test across different labs. In the general clinical population there are high false-positive and false-negative rates for the HerceptTest as well as the fluorescent in situ hybridization (FISH) test, which measures HER2 gene amplification. Although large central reference laboratories generally perform both these tests well with low false-positive and false-negative rates, small-volume laboratories, particularly those that use homebrew tests, have very high false-positive and false-negative rates. Discussions regarding the interpretation of the IHC test, as well as whether it is the most accurate test to use (i.e., compared with FISH testing for HER2 gene amplification) have also generated significant controversy.

SOURCES: Jacobs et al., 1999; Pauletti et al., 2000; Paik et al., 2002; Ellis et al., 2005; IOM, 2006; Perez et al., 2006; Reddy et al., 2006.

problems. In the UK HER2 Quality Assurance Program, which publishes its collective results (Rhodes et al., 2004), retesting of over 100 European laboratories on 6 successive occasions resulted, over a 2-year period, in a significant improvement in the number of laboratories achieving acceptable HER2 test results. The U.S. National Comprehensive Cancer Network HER2 Task Force recently recommended that "HER2 testing should be done only in laboratories accredited to perform such testing," noting that "such proficiency testing will probably become mandatory for laboratory accreditation in the future" (IOM, 2006; Carlson et al., 2006).

A SPECIAL CHALLENGE OF PHARMACOGENOMICS— CODEVELOPING DIAGNOSTIC-THERAPEUTIC COMBINATIONS

Because aberrant cell growth is a hallmark of cancers, oncology drug development has traditionally focused on agents that inhibit the basic machinery of cell division. As a result, these drugs often have significant side effects due to activity against normal proliferating tissues in the body. Many new cancer therapies are being developed to target the specific molecular changes in cancer cells that allow them to bypass normal regulation of signaling pathways that control cell growth and survival, with the goal of greater efficacy and fewer side effects. However, because of the heterogeneity among tumors, it is important to develop accompanying diagnostic tests that can identify those patients with the specific molecular changes targeted by a drug, who are most likely to benefit from that particular drug. Yet development of biomarker-based tests has lagged and is often undertaken outside the company developing the drug.

A prime example of this phenomenon is the development process for the targeted drug trastuzumab (Herceptin) and the accompanying diagnostic, the HercepTest (Box 3-6). In that case, although a biomarker test used by Genentech to select patients for inclusion in the clinical trials of trastuzumab was invaluable for successfully demonstrating efficacy of the drug, the test was deemed inadequate for commercial clinical use. As a result, another company (Dako Corporation) was asked to develop a commercial test at a very late stage of the drug development process. In the case of the epidermal growth factor receptor (EGFR) inhibitor cetuximab, a biomarker test to measure EFGR expression was used to select patients for clinical trials and a commercial test was again developed by Dako, but the evaluation of EGFR expression by immunohistochemistry has since been shown to be invalid for selecting responders, and a replacement test has yet to be proposed or developed (Box 3-7). It has been suggested that a diagnostic marker for EGFR inhibitors will need to incorporate various elements of the many signaling pathways that lie downstream from EGFR (Grunwald and Hidalgo, 2003). Another example is the development of antiestrogen therapies and the estrogen receptor biomarkers tests, which were completely separate in time and place (Box 3-5).

As noted in Chapter 2, the expectations of a biomarker test with respect to accuracy and performance vary depending on how it is used. A pharmaceutical company may have sufficient confidence in a biomarker they apply in phase I to establish a dose for a phase II trial even if the biomarker has not undergone stringent clinical qualification; the risk or benefit is theirs. But in the clinic, a responder/nonresponder stratification biomarker test that is going to be used to determine the appropriate treatment plan for individual patients must be highly accurate. Otherwise, a large number of patients could miss an opportunity for beneficial, life-saving therapy, while others could undergo expensive treatments and endure side effects with no chance of benefit.

Although patient stratification biomarkers are at the heart of personalized medicine, they can also create a conundrum for industry that does not arise with other biomarker applications. For example, few would argue that biomarkers that streamline the drug development process by facilitating earlier elimination of drug leads that are destined to fail or that improve dose selection should not be used. Similarly, even pharmaceutical marketing groups would support use of patient stratification biomarkers if the responder population was so small as to make stratification markers essential to demonstrate efficacy and FDA approval, as was the case for trastuzumab. But if a pharmaceutical company gets FDA approval for a drug without the use of stratification markers, it may debate whether to develop a biomarker test for patient stratification even if only one-fourth of patients respond well, as this is likely to limit the number of patients who take the drug (IOM, 2006).

Could progress in understanding cancers that enables cancer classification and thus patient stratification based on biomarker tests even lead to disincentives to drug development? For example, if the approximately 150,000 people diagnosed with colon cancer each year in the United States could be divided into multiple subsets, each with a different targeted therapy, then the market for any single drug is significantly reduced and companies will have less opportunity to recoup their developmental costs. In other words, biomarker tumor classification could essentially convert common cancers into orphan diseases, because the number of patients with any single particular subtype would be too small for companies to justify the enormous investment needed to develop novel drugs (Rawson, 2006).

Drugs that target the EGFR again provide a case in point. Four EGFR inhibitors have been approved by the FDA to treat patients with specific types of cancer, and several more are in development (Grunwald and Hidalgo, 2003). However, to date, there are no valid biomarker tests to accurately predict which patients are most likely to respond to each drug (Box 3-7). Such biomarker tests would be enormously helpful to clinicians and patients who must make treatment choices, but the individual sponsoring drug companies historically have not had sufficient incentives to discover stratification markers, nor the expertise to do so and develop the markers as diagnostic tests.

Industry perspectives are slowly changing regarding the strategic value of stratification biomarkers, and some companies are now devoting considerable resources to discovering stratification markers and working with diagnostics companies to convert these into molecular diagnostic tests. They are appreciating that patient stratification is both better medicine and better business, even when stratification is not essential for approval of the drug. First, health care resources are limited, and they risk unfavorable reimbursement decisions if expensive cancer drugs are effective for only a small fraction of the patients. In Great Britain and other countries with government-funded medicine, cost-effectiveness is considered in making coverage decisions (IOM, 2006; see Chapter 4). The drug company can either lower its price or stratify the patients to increase the cost-effectiveness. In essence, they may achieve a higher price if they can direct their therapy to those patients who will benefit. Second, companies also realize that if they do not stratify their patients, their competitor may do so and rapidly take over the market share. Finally, the corollary to identifying the responder population is identification of a nonresponder group for which appropriate therapy can then be developed. Genentech and other companies are now devoting considerable efforts to biomarker development aimed at patient stratification (Waring, 2006). But new strategies, methods, and infrastructure are needed to leverage and integrate the available data to better inform the biology, as noted in Chapter 2.

Public funding is also being directed toward filling the stratification biomarker gap. For example, the new NCI Lung Cancer Program has announced that it will undertake a clinical trial that will attempt to define

BOX 3-7   EGFR Inhibitors—
The Quest for Targeting Biomarkers

Most clinical trials conducted with EGFR inhibitors have not selected patients on the basis of a specific molecular marker. This is perhaps not surprising, given that EGFR overexpression is not tightly correlated with cancer progression, in contrast to HER2 expression in breast cancer. Available data are inadequate to determine what biomarkers might reliably indicate EGFR dependence and thereby select specific subsets of patients for treatment.

Gefitinib and Erlotinib

Gefitinib, owned by AstaZeneca, and erlotinib, marketed by OSI Pharmaceuticals in partnership with Genentech and Roche, are two examples of small molecule drugs that entered clinical trials without the use of a biomarker. A small clinical trial of gefitinib demonstrated a 10 percent response rate in patients with lung cancer, and the FDA granted accelerated approval in 2003. However, in December 2004, the FDA released a statement notifying of the failure of a large clinical trial of gefitinib to show an overall survival advantage compared with placebo in treating patients with lung cancer. In June 2005, FDA issued a new label for gefitinib "that limits use to patients with cancer who in the opinion of their treating physician, are currently benefiting, or have previously benefited, from gefitinib treatment." Nonetheless, researchers express optimism for the drug if appropriate biomarkers can be identified and validated for selecting a responsive patient population. Research shows that some patients who respond to gefitinib have amplifications and/or mutations in the EGFR gene, although response to treatment was quite variable, even for the same EGFR mutation type.

Erlotinib, approved by the FDA in 2004, showed an average two-month survival benefit for patients with nonsmall cell lung cancer compared with placebo. Further analysis showed that survival benefit correlated with EGFR status. In about one-third of the patients, tumor cells were examined to see whether they had high or low levels of EGFR. Among the approximately 55 percent who had high EGFR expression, the effect on survival was much greater than it was in people whose EGFR levels were low.

In both of these cases, the targets of the aforementioned EGFR inhibitors were not validated and biomarkers were not used to assess whether the drugs were actually working for patients. The result has been a lack of empirical, clinically derived data and inefficient adoption into clinical practice.

Cetuximab

Cetuximab, a monoclonal antibody made by ImClone, was approved by the FDA in 2004. Approval was based on a clinical trial that used an immunohistochemisty test for EGFR expression (EGFR pharmDx made by Dako and approved simultaneously by the FDA in 2004) to select colorectal cancer patients likely to respond to cetuximab. Patients were not entered into the clinical trials of cetuximab unless they had a positive result in the EGFR test (i.e., 1 percent or greater tumor cells showing positivity). The tumor response rate was 22.9 percent in patients who received cetuximab in combination with irinotecan, and 10.8 percent in patients who received cetuximab alone. However, no trials were performed with EGFR-negative patients, and further evaluation has shown that therapeutic response does not correlate with EGFR positivity, either by the number of positive cells or by staining intensity, perhaps because the staining pattern for EGFR is often quite heterogeneous. In March 2005, Chung et al. reported that EGFR-negative colorectal cancer patients treated with cetuximab in a nonstudy setting had a 25 percent response rate, suggesting that exclusion of patients from cetuximab treatment based on EGFR status is unwarranted. Thus, the EGFR test may have increased the probability that cetuximab would be approved, but it is not a valid test for making treatment decisions in the clinic. However, the FDA-approved drug label still specifies that it be used for the treatment of EGFR-expressing colorectal cancer.

Panitumumab

Panitumumab, a fully humanized monoclonal antibody against EGFR made by Amgen Inc., received accelerated FDA approval in September 2006 for treatment of EGFR-expressing metastatic colorectal cancer in patients with progression following chemotherapy. A randomized controlled trial of 463 patients demonstrated a significant improvement in progression-free survival in patients receiving panitumumab (mean of 96 days versus 60 days for patients receiving best supportive care). There was no difference in overall survival however, and the approval stipulates that the manufacturer must conduct a postmarketing trial to determine whether the drug improves survival in patients with fewer prior chemotherapies. Enrollment in the phase III trial was limited to patients whose tumors were positive for EGFR expression, defined as at least 1+ membrane staining in > 1 percent of tumor cells by the Dako EGFR pharmDx test kit (approved by FDA in September 2006 to assess patient eligibility for panitumumab as well as cetuximab). The majority of patients' tumors exhibited EGFR expression in 10 percent or

continued

BOX 3-7 Continued

more of tumor cells, with no evidence of a correlation between either the proportion of cells expressing EGFR or the intensity of EGFR expression.

SOURCES: FDA News, 2003; FDA, 2004a,c; FDA News, 2004a,b; Miller, 2004; Chung et al., 2005; Hirsch and Witta, 2005; Takano et al., 2005; Amgen, 2006; FDA News, 2006b; Hsieh et al., 2006.

---

a panel of genomic and proteomic pharmacodynamic markers to predict response to EGFR inhibitors in patients with nonsmall cell lung cancer. The trial will be supported by funds from the NCI director's discretionary budget reserve, and it will be conducted in conjunction with the FDA and CMS (Goldberg and Golderg, 2006; Niederhuber, 2006).

Progress in this field could be accelerated by better coordinating the development of biomarker diagnostics and new drugs. Such coordinated development could help companies choose the most promising drug leads, optimize clinical trial designs, and facilitate rapid and effective adoption into clinical practice (FDA, 2004b). However, there are many challenges to be addressed before this ideal approach becomes reality (IOM, 2006). For example, the cost and risks of diagnostic development are significant when clinical validity and utility must be established, and they add substantially to the existing high cost of drug development (estimated at $400–800 million, on average (Frank, 2003)). Companies may be unwilling to take the risk of investing in diagnostic development in the earlier phases of drug development, when approval of the drug is so uncertain. (On average, only 1 out of 5 Investigational New Drugs achieves FDA approval; Dimasi, 2001). But timing is key for the coapproval and marketing of drug-diagnostic combinations. Companies need to find better ways to integrate basic and clinical research efforts and emphasize the search for subpopulations based on theoretical and empirical evidence prior to phase III to avoid the rush near end of drug development (i.e., immediately prior to drug approval) to develop and validate the accompanying diagnostic.

Strategies to minimize the costs of diagnostics development and to facilitate risk sharing between pharmaceutical and diagnostics companies would also encourage development efforts. One possibility would be to link FDA approvals of therapeutics and the associated response-predicting diagnostics, such that one is contingent on the other. For example, one possible approach might be to provide contingent FDA approval of a drug by requiring postapproval reporting on diagnostic performance and subsequent submission of a PMA or 510(k) application for the diagnostic (IOM, 2006; Lipshutz, 2006). However, it is not clear that the FDA could compel a diagnostics company to sponsor a submission when the drug is sponsored by an unrelated pharmaceutical company. Furthermore, it seems unlikely that the FDA would rescind approval for a drug if the biomarker is subsequently shown to be invalid, as in the case of cetuximab.

The FDA should more clearly delineate the expectations and requirements for approval of diagnostic-therapeutic combinations. The FDA's "Critical Path" white paper placed high importance on personalized medicine and the codevelopment of diagnostics and therapeutics, noting that new trial designs and methods are needed, but it did not lay out specific plans for how to how to facilitate codevelopment (FDA, 2004b). In its April 2005 concept paper on codevelopment, the FDA noted that codevelopment applies when the use of an in vitro diagnostic is mandatory for drug selection for patients, or when optional use during drug development may assist in understanding disease mechanisms and in selecting clinical trial populations. Furthermore, codevelopment applies to a device-drug combination product, as well as to in vitro devices and drugs sold separately. The concept paper explicitly stated that drug selection biomarkers, particularly for high-risk conditions, were expected to be subject to PMA reviews (FDA, 2005b). In response, industry representatives expressed concern that the paper proposed higher hurdles for diagnostic approval than current requirements and that clinical utility is not explicitly defined in the act (Hinman et al., 2006). A new guidance document specifically focused on diagnostic-therapeutic combinations is being drafted by the FDA, taking into account feedback on the concept paper (Woodcock, 2006), but the content, impact, and enforceability are unknown at this time.

Because more than one FDA center will often be involved in the approval or clearance decisions in the case of diagnostic-therapeutic combinations, the agency should also clarify the roles of each center and focus on ensuring coordination between the centers to facilitate clearance or approval of molecular diagnostics. In addition, the FDA needs more dynamic ways of changing a drug's label when new data for selecting appropriate target populations emerge. When a biomarker test linked to a drug is found to be invalid (as in the case of cetuximab), the FDA should move quickly to make the necessary label changes. Conversely, when new biomarkers are found to aid therapeutic decisions for existing drugs, a formal mechanism is needed to evaluate the evidence and consider appropriate label changes.

SUMMARY AND CONCLUSIONS

The standards used to demonstrate the validity of biomarkers vary considerably, in part because there is no overarching leadership in the field to set uniform consensus standards for biomarker development. The FDA and CMS have some authority over diagnostic tests, but oversight has been variable and unpredictable and, in many cases, inadequate to ensure the safety, effectiveness, and value of tests on the market. Oversight by federal agencies has been evolving recently, and FDA in particular has taken some positive initial steps, but there is still a need for clarification, uniformity, and leadership in this area. The process of biomarker development and evaluation could be improved by making it more transparent, consistent, and effective.

First, government agencies, including NIH, the FDA, CMS, and NIST, and non-government stakeholders, including academia, the pharmaceutical and diagnostics industry, and health care payors, should work together to develop a transparent process for creating well-defined consensus standards and guidelines for biomarker development, validation, qualification, and use to reduce the uncertainty in the process of development and adoption. NIST or another appropriate federal agency should provide a leadership role in the process, coordinating and overseeing interagency activities.

Second, the FDA should clarify its authority over biomarker tests linked to clinical decision making and then establish and consistently apply clear guidelines for the oversight of those tests. In addition, the appropriate federal agency (e.g., the FDA or the FTC) should monitor and enforce marketing claims made about molecular diagnostics. Variability and unpredictability in oversight can reduce interest and investment in developing innovative diagnostics, while inadequate evaluation and oversight could lead to harm for patients and unnecessary costs for society.

Third, the FDA and industry should work together to facilitate the codevelopment of diagnostic-therapeutic combinations. The FDA should more clearly delineate the expectations and requirements for diagnostic-therapeutic combination approval, and companies need to better integrate basic and clinical research rather than waiting to contract biomarker development in the late stages of phase III testing. Coordinated development of diagnostics and therapeutics is an important component in the quest for personalized medicine; it could help companies choose the most promising drug leads, optimize clinical trial designs, and facilitate rapid and effective adoption into clinical practice.

Finally, CMS should develop a specialty area for molecular diagnostics under CLIA. In contrast to other high-complexity tests, CMS has not created a specialty area for molecular diagnostics that could mandate, among other requirements, participation in specified proficiency testing programs. The minimum generic standards set by CMS under CLIA are inadequate to ensure high-quality, accurate test results.

REFERENCES

Amgen. 2006. *Vectibix (panitumumab).* [Online]. Available: http://wwwext.amgen.com/pdfs/products/vectibix_pi.pdf [accessed October 2006].

Barker PE. 2003. Cancer biomarker validation: Standards and process: Roles for the National Institute of Standards and Technology (NIST). *Annals of the New York Academy of Sciences* 983:142-150.

Barker PE, Wagner PD, Stein SE, Bunk DM, Srivastava S, Omenn GS. 2006. Standards for plasma and serum proteomics in early cancer detection: a needs assessment report from the National Institute of Standards and Technology–National Cancer Institute Standards, Methods, Assays, Reagents and Technologies Workshop, August 18–19, 2005. *Clinical Chemistry* 52(9):1669-1674.

Bast RC Jr, Ravdin P, Hayes DF, Bates S, Fritsche H Jr, Jessup JM, Kemeny N, Locker GY, Mennel RG, Somerfield MR. 2001. 2000 update of recommendations for the use of tumor markers in breast and colorectal cancer: Clinical practice guidelines of the American Society of Clinical Oncology. *Journal of Clinical Oncology* 19(6):1865-1878.

Beavis R. 2005. The Paris consensus. *Journal of Proteome Research* 4(5):1475.

Bossuyt PM, Reitsma JB, Bruns DE, Gatsonis CA, Glasziou PP, Irwig LM, Lijmer JG, Moher D, Rennie D, de Vet HC. 2003. Towards complete and accurate reporting of studies of diagnostic accuracy: The STARD Initiative. *Annals of Internal Medicine* 138(1):40-44.

Brazma A, Hingamp P, Quackenbush J, Sherlock G, Spellman P, Stoeckert C, Aach J, Ansorge W, Ball CA, Causton HC, Gaasterland T, Glenisson P, Holstege FC, Kim IF, Markowitz V, Matese JC, Parkinson H, Robinson A, Sarkans U, Schulze-Kremer S, Stewart J, Taylor R, Vilo J, Vingron M. 2001. Minimum information about a microarray experiment (MIAME)—Toward standards for microarray data. *Nature Genetics* 29(4):365-371.

Carlson RW, Moench SJ, Hammond ME, Perez EA, Burstein HJ, Allred DC, Vogel CL, Goldstein LJ, Somlo G, Gradishar WJ, Hudis CA, Jahanzeb M, Stark A, Wolff AC, Press MF, Winer EP, Paik S, Ljung BM. 2006. HER2 testing in breast cancer: NCCN Task Force report and recommendations. *Journal of the National Comprehensive Cancer Network* 4 (Suppl 3):S1-S22; quiz S23-S4.

CDC (Centers for Disease Control and Prevention). 2006. *Evaluation of Genomic Applications in Practice and Prevention (EGAPP): Implementation and Evaluation of a Model Approach.* [Online]. Available: http://www.cdc.gov/genomics/gtesting/ACCE/fbr.htm [accessed June 2006].

Chung KY, Shia J, Kemeny NE, Shah M, Schwartz GK, Tse A, Hamilton A, Pan D, Schrag D, Schwartz L, Klimstra DS, Fridman D, Kelsen DP, Saltz LB. 2005. Cetuximab shows activity in colorectal cancer patients with tumors that do not express the epidermal growth factor receptor by immunohistochemistry. *Journal of Clinical Oncology* 23(9):1803-1810.

CMS (Centers for Medicare & Medicaid Services). 2005. *Interpretive Guidelines for Laboratories.* [Online]. Available: http://www.cms.hhs.gov/CLIA/03_Interpretive_Guidelines_for_Laboratories.asp [accessed July 2006].

———. 2006. *CLIA Interpretive Guidelines for Laboratories, Subpart K—Quality System for Nonwaived Testing.* [Online]. Available: http://www.cms.hhs.gov/CLIA/downloads/apcsubk1.pdf [accessed May 2006].

CMS, DHHS (Department of Health and Human Services). 2004. CLIA Laboratory Requirements. 42 CFR Part 493.

Cottingham K. 2005. Universal proteomics guidelines debated. *Journal of Proteome Research* 4(4):1051.

Couzin J. 2006. Genomics. Microarray data reproduced, but some concerns remain. *Science.* 313(5793):1559.

DHHS (Department of Health and Human Services). 1988. H.R. Rep. No. 100-899, at 28.

———. 1999. Public Health Service Act. 42 USC 263a:353.

———. 2003. Medicare, Medicaid, and CLIA programs; Laboratory requirements relating to quality systems and certain personnel qualifications; Final rule. *Federal Register* 68(16):3698.

———. 2006. *Federal Register* 71(77):20707-20708.

DHHS, FDA (Food and Drug Administration). 2006. *Critical Path Opportunities List.* [Online]. Available: http://www.fda.gov/oc/initiatives/criticalpath/reports/opp_list.pdf [accessed August 2006].

Diaz LK, Sneige N. 2005. Estrogen receptor analysis for breast cancer: current issues and keys to increasing testing accuracy. *Advances in Anatomic Pathology* 12(1):10-19.

Dimasi JA. 2001. Risks in new drug development: approval success rates for investigational drugs. *Clinical Pharmacology and Therapeutics* 69(5):297-307.

Ellis CM, Dyson MJ, Stephenson TJ, Maltby EL. 2005. HER2 amplification status in breast cancer: A comparison between immunohistochemical staining and fluorescence in situ hybridisation using manual and automated quantitative image analysis scoring techniques. *Journal of Clinical Pathology* 58(7):710-714.

Ellis IO, Bartlett J, Dowsett M, Humphreys S, Jasani B, Miller K, Pinder SE, Rhodes A, Walker R. 2004. Best practice no 176: Updated recommendations for HER2 testing in the UK. *Journal of Clinical Pathology* 57(3):233-237.

FDA (Food and Drug Administration). 2003a. *Letter from OIVD to Roche Molecular Diagnostics Re: AmpliChip.* [Online]. Available: http://www.fda.gov/cdrh/oivd/amplichip.html [accessed July 2006].

————. February 26, 2003b. *CDRH, OVID, Analyte Specific Reagents, Small Entity Compliance Guidance, Guidance for Industry.* [Online]. Available: http://www.fda.gov/cdrh/oivd/guidance/1205.html [accessed July 2006].

————. February 2004a. *New Device Approval: DakoCytomation EGFR pharmDx-P030044.* [Online] Available: http://www.fda.gov/cdrh/mda/docs/p030044.html [Accessed August 2006].

————. July 2004b. *Letter to Correlogic Systems, Inc.* [Online]. Available: http://www.fda.gov/cdrh/oivd/letters/071204-correlogic.html [accessed July 2006].

————. December 2004c. *FDA Statement on Iressa.* [Online]. Available: http://www.fda.gov/bbs/topics/news/2004/new01145.html [accessed July 2006].

————. March 2005a. *Guidance for Industry: Pharmacogenomic Data Submissions.* [Online]. Available: http://www.fda.gov/CbER/gdlns/pharmdtasub.htm [accessed July 2006].

————. April 2005b. *Drug-Diagnostic Co-Development Concept Paper.* [Online]. Available: http://www.fda.gov/cder/genomics/pharmacoconceptfn.pdf [accessed September 2006].

————. August 2005c. *Class II Special Controls Guidance Document: RNA Preanalytical Systems (RNA collection, stabilization and purification systems for RT-PCR used in molecular diagnostic testing.* [Online]. Available: http://www.fda.gov/cdrh/oivd/guidance/1563.html [accessed July 2006].

————. August 2005d. *Letter to Nanogen Corporation.* [Online]. Available: http://www.fda.gov/cdrh/oivd/letters/081105-nanogen.html [accessed July 2006].

————. August 2005e. *Warning Letter to Access Genetics.* [Online]. Available: http://www.fda.gov/cdrh/oivd/letters/080105-access.html [accessed July 2006].

————. 2006a. *CDRH (Center for Devices and Radiological Health) Search Guidance Database.* [Online]. Available: http://www.accessdata.fda.gov/scripts/cdrh/cfdocs/cfgpp/search.cfm [accessed July 2006].

————. 2006b. *CLIA—Clinical Laboratory Improvement Amendments.* [Online]. Available: www.fda.gov/cdrh/clia/ [accessed July 2006].

————. 2006c. *FDA history.* [Online]. Available: http://www.fda.gov/oc/history/ [accessed July 2006].

————. 2006d. *MicroArray Quality Control (MAQC) Project.* [Online]. Available: http://www.fda.gov/nctr/science/centers/toxicoinformatics/maqc/ [accessed August 2006].

FDA News. May 5, 2003. *FDA Approves New Type of Drug for Lung Cancer.* [Online]. Available: http://www.fda.gov/bbs/topics/NEWS/2003/NEW00901.html [accessed August 2006].

————. February 12, 2004a. *FDA Approves Erbitux for Colorectal Cancer.* [Online]. Available: http://www.fda.gov/bbs/topics/NEWS/2004/NEW01024.html [accessed August 2006].

————. November 19, 2004b. *FDA Approves New Drug for the Most Common Type of Lung Cancer.* [Online]. Available: http://www.fda.gov/bbs/topics/news/2004/NEW01139.html [accessed August 2006].

————. 2006a. *FDA Drafts Regulatory Direction to Industry for Active Ingredients Used in Medical Tests.* [Online]. Available: http://www.fda.gov/bbs/topics/NEWS/2006/NEW01444.html [accessed September 2006].

———. September 27, 2006b. *FDA Approves New Drug for Colorectal Cancer, Vectibix.* [Online]. Available: http://www.fda.gov/bbs/topics/NEWS/2006/NEW01468.html [accessed October 19, 2006].

Fitzgibbons PL, Page DL, Weaver D, Thor AD, Allred DC, Clark GM, Ruby SG, O'Malley F, Simpson JF, Connolly JL, Hayes DF, Edge SB, Lichter A, Schnitt SJ. 2000. Prognostic factors in breast cancer. College of American Pathologists Consensus Statement 1999. *Archives of Pathology and Laboratory Medicine* 124(7):966-978.

Frank RG. 2003. New estimates of drug development costs. *Journal of Health Economics* 22(2):325-330.

Frueh FW. 2006. Impact of microarray data quality on genomic data submissions to the FDA. *Nature Biotechnology* 24(9):1105-1107.

FTC (Federal Trade Commission). 2006. *At-Home Genetic Tests: A Healthy Dose of Skepticism May Be The Best Prescription.* [Online]. Available: http://www.ftc.gov/bcp/edu/pubs/consumer/health/hea02.htm [accessed July 2006].

Genetics & Public Policy Center. 2006. *News Releases: Lax Oversight of Genetic Tests "A Risk to Public Health"—Public Policy Groups File Petition for Rulemaking with CMS.* [Online]. Available: http://www.dnapolicy.org/news.release.php?action=detail&pressrelease_id=61 [accessed October 2006].

Goldberg K, Goldberg P, eds. 2006. *The Cancer Letter* 32(24).

Grunwald V, Hidalgo M. 2003. Developing inhibitors of the epidermal growth factor receptor for cancer treatment. *Journal of the National Cancer Institute* 95(12):851-867.

Gutman S. June 19, 2000. Presentation at the meeting of the IOM Committee on Technologies for the Early Detection of Breast Cancer. Washington, DC.

Hackett JL, Gutman SI. 2005. Introduction to the Food and Drug Administration (FDA) regulatory process. *Journal of Proteome Research* 4(4):1110-1113.

Hammond ME, Fitzgibbons PL, Compton CC, Grignon DJ, Page DL, Fielding LP, Bostwick D, Pajak TF. 2000. College of American Pathologists Conference XXXV: Solid tumor prognostic factors—Which, how and so what? Summary document and recommendations for implementation. Cancer Committee and Conference Participants. *Archives of Pathology and Laboratory Medicine* 124(7):958-965.

Hayes DF, Bast RC, Desch CE, Fritsche H Jr, Kemeny NE, Jessup JM, Locker GY, Macdonald JS, Mennel RG, Norton L, Ravdin P, Taube S, Winn RJ. 1996. Tumor marker utility grading system: A framework to evaluate clinical utility of tumor markers. *Journal of the National Cancer Institute* 88(20):1456-1466.

Hayes DF, Trock B, Harris AL. 1998. Assessing the clinical impact of prognostic factors: When is "statistically significant" clinically useful? *Breast Cancer Research and Treatment* 52(1-3):305-319.

Heller M. March 21, 2006. The basics and direction of the regulation of molecular diagnostic and prognostic devices. Presentation at the IOM workshop on Developing Biomarker-based Tools for Cancer Screening, Diagnosis, and Treatment. Washington, DC.

Hinman LM, Huang SM, Hackett J, Koch WH, Love PY, Pennello G, Torres-Cabassa A, Webster C. 2006. The drug diagnostic co-development concept paper Commentary from the 3rd FDA-DIA-PWG-PhRMA-BIO Pharmacogenomics Workshop. *The Pharmacogenomics Journal* 6(6):375-380.

Hirsch FR, Witta S. 2005. Biomarkers for prediction of sensitivity to EGFR inhibitors in non-small cell lung cancer. *Current Opinion in Oncology* 17(2):118-1122.

Hsieh MH, Fang YF, Chang WC, Kuo HP, Lin SY, Liu HP, Liu CL, Chen HC, Ku YC, Chen YT, Chang YH, Chen YT, Hsi BL, Tsai SF, Huang SF. 2006. Complex mutation patterns of epidermal growth factor receptor gene associated with variable responses to gefitinib treatment in patients with non-small cell lung cancer. *Lung Cancer* 53(5):311-322.

Hudson K, Javitt GH. 2006. Federal neglect: Regulation of genetic testing. *Issues in Science and Technology* 22(3):59-66.

Hudson KL. 2006. Genetic testing oversight. *Science* 313(5795):1853.

Hudson KL, Murphy JA, Kaufman DJ, Javitt GH, Katsanis SH, Scott J. 2006. Oversight of U.S. genetic testing laboratories. *Nature Biotechnology* 24(9):1083-1090.

HUPO (Human Proteome Organization). 2006. *HUPO Proteomics Standards Initiative.* [Online]. Available: http://psidev.sourceforge.net/ [accessed September 2006].

IOM (Institute of Medicine). 2005. *Saving Women's Lives.* Joy JE, Penhoet EE, Petitti DB, eds. Washington, DC: The National Academies Press.

———. 2006. *Developing Biomarker-based Tools for Cancer Screening, Diagnosis, and Treatment: The State of the Science, Evaluation, Implementation, and Economics. A Workshop.* Patlak M, Nass S, rapporteurs. Washington, DC: The National Academies Press.

Jacobs TW, Gown AM, Yaziji H, Barnes MJ, Schnitt SJ. 1999. Specificity of HercepTest in determining HER-2/neu status of breast cancers using the United States Food and Drug Administration-approved scoring system. *Journal of Clinical Oncology* 17(7):1983-1987.

Javitt G. 2006. *Institute of Medicine Committee on Developing Cancer Biomarkers.* Presentation at the meeting of the Committee on Developing Cancer Biomarkers, Meeting 2. Washington, DC.

Lipshutz, R. 2006. Coordinating the development of biomarkers and targeted therapies: A diagnostics industry perspective. Presentation at the IOM workshop on Developing Biomarker-based Tools for Cancer Screening, Diagnosis, and Treatment. Washington, DC.

McGuire WL. 1975. Current status of estrogen receptors in human breast cancer. *Cancer* 36(2):638-644.

McShane LM, Altman DG, Sauerbrei W, Taube SE, Gion M, Clark GM. 2005. Reporting recommendations for tumor marker prognostic studies. *Journal of Clinical Oncology* 23(36):9067-9072.

MGED (Microarray Gene Expression Data). 2006. *MGED Workgroups.* [Online]. Available: http://www.mged.org/Workgroups/index.html [accessed July 2006].

MGED Society. September 2005. *MIAME Checklist.* [Online]. Available: http://www.mged.org/Workgroups/MIAME/miame_checklist.html [accessed June 2006].

Miller RT. 2004. Epidermal growth factor receptor and Erbitux. *Pro Path Immunohistochemistry.*

NCI Division of Cancer Prevention. 2005. *The Early Detection Research Network: Translational Research to Identify Early Cancer and Cancer Risk.* 3rd edition. DHHS. [Online]. Available: http://edrn.nci.nih.gov/docs/progress-reports/edbi6.pdf/download [accessed July 2006].

———. 2006. *NIST-EDRN Workshop on Standards and Metrology for Cancer Diagnostics.* [Online]. Available: http://www.cancer.gov/prevention/cbrg/edrn/workshop/index.html [accessed August 2006].

Niederhuber JE. 2006. Director's update: New focus on lung cancer research. *NCI Cancer Bulletin* 3(21).

NIH. 2006. *NIH Roadmap for Medical Research: Standards in Proteomics.* [Online]. Available: http://nihroadmap.nih.gov/buildingblocks/proteomics/ [accessed July 2006].

OECD (Organisation for Economic Co-operation and Development). July 2006. *Draft Guidelines for Quality Assurance in Molecular Genetic Testing.* [Online]. Available: http://www.oecd.org/dataoecd/43/26/37103271.pdf#search=%22quality%20genetic%20site%3Aoecd.org%22 [accessed October 2006].

Paik S, Bryant J, Tan-Chiu E, Romond E, Hiller W, Park K, Brown A, Yothers G, Anderson S, Smith R, Wickerham DL, Wolmark N. 2002. Real-world performance of HER2 testing—National Surgical Adjuvant Breast and Bowel Project experience. *Journal of the National Cancer Institute* 94(11):852-854.

Pauletti G, Dandekar S, Rong H, Ramos L, Peng H, Seshadri R, Slamon DJ. 2000. Assessment of methods for tissue-based detection of the HER-2/neu alteration in human breast cancer: A direct comparison of fluorescence in situ hybridization and immunohistochemistry. *Journal of Clinical Oncology* 18(21):3651-3664.

Perez EA, Suman VJ, Davidson NE, Martino S, Kaufman PA, Lingle WL, Flynn PJ, Ingle JN, Visscher D, Jenkins RB. 2006. HER2 testing by local, central, and reference laboratories in specimens from the North Central Cancer Treatment Group N9831 intergroup adjuvant trial. *Journal of Clinical Oncology* 24(19):3032-3038.

Perkel JM. 2006a. In search of microarray standards. *The Scientist* 20(4):78.

Perkel JM. 2006b. Six things you won't find in the MAQC. *The Scientist* 20(11):68.

Rawson K. 2006. Getting personal: FDA's plan to save the drug industry. *The RPM Report* 1(9).

Reddy JC, Reimann JD, Anderson SM, Klein PM. 2006. Concordance between central and local laboratory HER2 testing from a community-based clinical study. *Clinical Breast Cancer* 7(2):153-157.

Rhodes A, Borthwick D, Sykes R, Al-Sam S, Paradiso A. 2004. The use of cell line standards to reduce HER-2/neu assay variation in multiple European cancer centers and the potential of automated image analysis to provide for more accurate cut points for predicting clinical response to trastuzumab. *American Journal of Clinical Pathology* 122(1):51-60.

Ross JS. 2005. Improving the accuracy of hormone receptor assays in breast cancer: An unmet medical need. *Future Oncology* 1(4):439-441.

Schmitt M, Harbeck N, Daidone MG, Brynner N, Duffy MJ, Foekens JA, Sweep FC. 2004. Identification, validation, and clinical implementation of tumor-associated biomarkers to improve therapy concepts, survival, and quality of life of cancer patients: Tasks of the Receptor and Biomarker Group of the European Organization for Research and Treatment of Cancer. *International Journal of Oncology* 25(5):1397-1406.

Shapiro JK, Prebula RJ. 2003. FDA's regulation of analyte-specific reagents. *Medical Device & Diagnostic Industry.* [Online]. Available: http://www.devicelink.com/mddi/archive/03/02/018.html [accessed August 2006].

Swanson BN. 2002. Delivery of high-quality biomarker assays. *Disease Markers* 18(2):47-56.

Takano T, Ohe Y, Sakamoto H, Tsuta K, Matsuno Y, Tateishi U, Yamamoto S, Nokihara H, Yamamoto N, Sekine I, Kunitoh H, Shibata T, Sakiyama T, Yoshida T, Tamura T. 2005. Epidermal growth factor receptor gene mutations and increased copy numbers predict gefitinib sensitivity in patients with recurrent non-small-cell lung cancer. *Journal of Clinical Oncology* 23(28):6829-6837.

Tan PK, Downey TJ, Spitznagel EL Jr, Xu P, Fu D, Dimitrov DS, Lempicki RA, Raaka BM, Cam MC. 2003. Evaluation of gene expression measurements from commercial microarray platforms. *Nucleic Acids Research* 31(19):5676-5684.

Waring P. March 20, 2006. Therapeutics industry perspectives/realities (examples of successes and difficulties/failures of targeted therapy). Presentation at the IOM workshop on Developing Biomarker-based Tools for Cancer Screening, Diagnosis, and Treatment. Washington, DC.

Woodcock J. Deputy Commissioner for Operations FDA. 2006. Personal communication.

4

Methods and Process Needed for Clinical Adoption and Evaluation of Biomarker-Based Diagnostics

In order for cancer biomarker tests to be used effectively in a clinical setting, their clinical risks and benefits must be assessed. Even for diagnostic tests that have received Food and Drug Administration (FDA) approval (which are few in number), the clinical utility has not been assessed. Clinical algorithms need to be developed that specify the target patient populations for the diagnostic test and the changes in patient management that follow from test results. Well-designed, prospective clinical studies are needed to demonstrate that the test results influence the patient's management such that clinical outcomes are improved. However, the studies necessary to develop evidence of the value of these tests may be costly and lengthy, especially for tests used for cancer screening. This deters diagnostic companies from conducting such studies. Instead they usually introduce biomarker-based products into the market via a 510(k) review process or by developing homebrew tests for in-house use (FDA, 2001; IOM, 2005b, see also Chapter 3).

These pathways to the market bypass the need to supply evidence of a test's clinical benefits and risks. They also do not require manufacturers to specify the patient population(s) for which the test should be used, or how the test fits into the clinical care pathway of a patient. Off-label use of FDA-approved biomarker tests also fosters clinical applications for purposes other than that for which they have been clinically validated. For example, most tests enter the market as cancer diagnostics, but they can then be used for cancer screening without adequate evaluation. The extension of the use of the prostate-specific antigen (PSA) test for prostate cancer from diagnosis to screening, described below, is a prime example of this scenario. Once adopted in such a fashion, it may be difficult or impossible to adequately assess the risks, benefits, and value of a screening test. Postmarket surveillance of diagnostic tests is minimal, and once insurers provide coverage for something, coverage is rarely withdrawn unless the item is removed from the market because of safety concerns (reviewed by IOM, 2001).

Ultimately, the value of a test to society also depends on its cost-effectiveness and economic impact. Although these factors have not generally been considered in coverage and adoption decisions for health care in the United States, interest in such assessments is increasing as the cost of medical care continues to rise. Few economic evaluations of diagnostics have been undertaken thus far, perhaps because of their relatively low cost compared with many drugs and other medical interventions (Rogowski, in press). However, the newer class of pharmacogenomics-based molecular diagnostics that will enable personalized medicine may come under closer scrutiny because of their potentially significant budget impact due to high drug costs and the high cost of adverse drug reactions. Nonetheless, economic evidence for this class of diagnostics is presently still quite limited (Rogowski, 2007).

This chapter provides an overview of the challenges and needs of technology assessment and adoption, with the goal of identifying possible ways to facilitate data collection and analysis to monitor and improve the value of biomarker tests. Examples described below, as well as in Chapter 3, illustrate the complexity of this topic. Problematic cases include instances in which markers approved for one purpose were widely diffused and adopted for another purpose without sufficient evidence; instances in which the use of markers for vital treatment guidance is based on small, poor-quality studies likely to be less than definitive; and many instances in which evidence for the value of markers to improve patient outcomes is flawed or insufficient.

THE CHALLENGE OF ASSESSING CLINICAL VALUE

As a result of the limited scope of FDA oversight of laboratory tests, biomarker tests often are applied in clinical settings with little assessment of their clinical utility for specific medical situations (Reid et al., 1995; Feinstein, 2002; Weinstein et al., 2005). This does not seem to hinder widespread clinical adoption, however.

Tests that are introduced for one indication may find much wider application in other settings. The blood test for prostate-specific antigen provides a telling example of rapid adoption of a test for a use that it was not approved for. The FDA approved the PSA test in 1985 for the detection of prostate cancer *recurrence*, but it is now widely used for prostate cancer *screening*. Most studies show that 50–60 percent of men (50 years and older) get recommended prostate cancer screening, and some advocates report that up to 75 percent of that target population undergoes regular screening for prostate cancer, despite the fact that the U.S. Preventive Services Task Force (USPSTF) gave it an "I" rating (Swan et al., 2003; Carlos et al., 2005). This rating indicates that the task force found insufficient evidence to give the PSA test its backing for prostate screening purposes, primarily because there was inconclusive evidence that early detection by screening improves health outcomes and substantial evidence of screening-related harms. The task force reviewed studies that suggested that preventing one death from prostate cancer in eight years would require annual PSA screening of about 1,000 men with the test (Research Triangle Institute, 2002; CDC, 2006). But substantial proportions of these men would be subject to such potential harms as false-positive tests, anxiety, and treatment-linked erectile dysfunction, incontinence, and bowel dysfunction.

The rapid adoption of the PSA test for prostate cancer screening illustrates the potential costs to both society and individuals of adopting a biomarker test before its clinical risks and benefits have been adequately assessed. One way to foster the key prospective studies needed for such assessments is to support them via government funds, public–private collaborations, or nonprofit consortia.

Such support was recently provided to fund prospective clinical trials of OncotypeDX and MammaPrint, two genomic tests for predicting the risk of breast cancer recurrence. Both tests use the gene expression signatures of breast tumors to determine which women with node-negative invasive breast cancers would be most likely to benefit from chemotherapy (Box 4-1). But there is a lack of studies that can firmly establish the clinical outcomes of using cancer biomarker tests. Well-designed prospective clinical studies of diagnostic tests are often lacking (IOM, 2006). In addition, clearly defined patient populations, relevant comparators, and intention-to-treat analyses of all participants by initial group assignment are often missing, nor do studies always have the long-term follow-up needed to adequately assess the health outcomes of a medical intervention. The end result is a lack of robust evidence of the effects of an intervention and how those effects compare with other interventions (IOM, 2006).

Two major disincentives for undertaking such studies are the cost and length of time needed to complete the study. As more biomarker tests are developed, it may become increasingly difficult to fund and undertake adequate studies to assess them all. For example, the cost of the trial to assess the OncotypeDX test (the Trial Assigning Individualized Options for Treatment, TAILORx) to NCI alone is estimated at $27 million for 5 years of the trial.[1] It could be argued that, in the long run, the cost of the trial will be small compared with unnecessarily treating many women with chemotherapy. However, technology is continually evolving as new discoveries are made, so by the time this study is finished, new data may indicate that a slightly different set of genes is even better at predicting outcomes. But without an infinite source of funding, the ability to launch additional studies will be limited.

EVIDENCE FOR COVERAGE

The lack of direct evidence for the value of diagnostics makes it difficult for insurers to make informed decisions about coverage for new tests. Performance characteristics of the test are often used to fill in a model of how the technology can detect a condition or change its management to give an improved health outcome (IOM, 2006). But such an approach can be too simplistic. For example, there is a test for variations in the gene that codes for the drug-metabolizing enzyme cytochrome P450 (CYP450). These variants can reduce or increase the enzyme's ability to metabolize certain drugs, including the anticoagulant warfarin. Therefore, a person who has a variant gene might benefit by lowering or raising the doses of those drugs. But other factors also can affect drug metabolism. These factors include other enzymes, coexisting disease, age, diet, and interactions with other drugs (reviewed by Takahashi and Echizen, 2003). Given this complex scenario, it is not yet clear how useful a test for just one influence on drug metabolism will be for patients who take warfarin. To address this question, the Critical Path Institute, a public–private partnership (see Chapter 2), has begun a randomized study of an individualized, genotype-based warfarin-dosing regimen versus standard care (Feigal, 2006).

---

[1]NCI has partnered with Genomic Health, Inc., in the TAILORx trial. This cost estimate does not include funds contributed by Genomic Health, Inc.

**BOX 4-1   Assessing the Value
of OncotypeDX and MammaPrint**

Chemotherapy is currently recommended for most women with node-negative breast cancer that is greater than 1 cm or has unfavorable pathology. But studies show that chemotherapy offers only a modest improvement in the 10-year survival rate, especially for women with estrogen receptor (ER)-positive disease treated with hormonal therapy. Many women could be spared the significant side effects of chemotherapy if there was a way to discern whether they have tumors not likely to be significantly affected by such toxic therapies, either because they are relatively indolent tumors not likely to recur and spread, and/or because they are relatively insensitive to the effects of chemotherapy.

Initial findings from studies on OncotypeDX suggest that this 21-gene test can predict the risk of recurrence for node-negative, ER-positive breast tumors. The studies identified a large subset of patients (about 50 percent) who were at very low risk of dying from breast cancer within 10 years. In one study, chemotherapy lowered the risk of recurrence by nearly 30 percent in women with a high recurrence score, but it reduced this risk by only about 1 percent in women with a low recurrence score.

These findings suggest that combining the oncotype recurrence score with tumor grade and size, or using it instead of these traditional prognostic factors, might help physicians better determine which women are at high risk of having a breast cancer recurrence and therefore might benefit from having more aggressive chemotherapy in addition to hormonal therapy. But none of these studies on OncotypeDX was a large, prospective, randomized clinical study that would be most likely to accurately assess the utility of the diagnostic in a clinical setting. The National Cancer Institute recently launched such a study, which is expected to last 10 years (with an additional follow-up of 20 years after initial therapies), and it will enroll over 10,000 women at 900 sites in the United States and Canada. The Trial Assigning Individualized Options for Treatment (TAILORx) is designed mainly to evaluate the effect of chemo-

---

Without evidence of clinically utility, many insurers are reluctant to cover the costs of innovative tests. Lack of coverage, in turn, often impedes their widespread adoption in clinical settings. But this poses a dilemma that was described in the Institute of Medicine report *Saving Women's Lives* (IOM, 2005b):

therapy (in addition to hormonal therapy) on women with ER-positive, node-negative breast cancers with recurrence scores in the intermediate range. These women will all receive hormonal therapy, but then they will be randomly assigned to receive chemotherapy or not in addition.

With support from the European Organisation for Research and Treatment of Cancer and an estimated €10M from Agendia, researchers are conducting another large, prospective, randomized clinical study of MammaPrint, a microarray test for a 70-gene expression signature that initial studies suggest is linked to breast cancer prognosis in women 60 years or younger with either ER-positive or ER-negative tumors. One study found that this gene signature outperformed traditional prognostic factors, such as tumor size and grade, in predicting recurrence within 10 years. To more fully assess this, the Microarray In Node-negative Disease may Avoid ChemoTherapy (MINDACT) study will randomly assign chemotherapy to half the women with breast cancers that appear to be at low risk of recurrence from their MammaPrint results, yet at high risk of recurrence based on traditional prognostic factors. Over 6,000 women will be followed for 6 years. The researchers will use the recurrence and survival rates for each treatment strategy to assess whether MammaPrint is more effective than standard prognostic factors in determining who will benefit the most from chemotherapy.

Although there is very little overlap in the genes assessed by these two tests, a recent study of 295 patient samples found highly concordant outcome predictions (about 80 percent) between the test results. This concordance likely occurs because the different gene sets reflect common cellular phenotypes and biological characteristics that are present in different groups of breast cancer patients, but the results do raise the question of whether biomarker tests should target genes that are at the origin of pathophysiological pathways, or the final genes that encode proteins that delineate the tumor phenotype.

SOURCES: Eifel et al., 2001; Goldhirsch et al., 2001; van de Vijver et al., 2002; Fisher et al., 2004; Paik et al., 2004, 2006; European Organisation for Research and Treatment of Cancer, 2005; Frantz, 2005; Fan et al., 2006; Habel et al., 2006; NCI, 2006.

. . . insurance coverage of the new technology would increase its use, providing both some of the resources needed for its developers to study its clinical value and more clinical experience with the new technology. Yet, once coverage is granted, there is little incentive (and more likely a disincentive) for companies to gather data and formally evaluate the clinical effectiveness of their new technology (p. 230).

Conditional coverage is one way to get around this dilemma. Conditional coverage by the Centers for Medicare & Medicaid Services (CMS) and other insurers could provide a means of collecting important data on the use, effectiveness, and value of biomarker tests before they are broadly adopted. Payors would agree to provisionally cover new tests with the proviso that, in the interim, data would be collected in conjunction with use of the test, to assess its clinical utility and value.

CMS has already used Coverage under Evidence Development (CED) for innovative diagnostic biomarker technologies, such as fluorodeoxy-glucose positron emission tomography (FDG-PET) scanning for cancer diagnosis, staging, and monitoring. CMS first determined that the evidence was not persuasive that FDG-PET scanning was a useful technology in all cancers. But based on some studies suggesting the usefulness of FDG-PET in certain cancers as a biomarker for cancer staging, diagnosis, and monitoring, the federal agency decided to provide coverage for such use of the imaging technology, dependent on the mandatory collection of clinical data (CMS, 2005a).

According to draft guidance put out by CMS (CMS, 2005a, 2006), the agency considers CED to be particularly useful in the following situations relevant to cancer biomarkers:

- To clarify the risks and benefits for off-label uses.
- To clarify the risks and benefits of a diagnostic or treatment in specific patient subgroups that other clinical trials have not addressed but that comprise a sizable portion of Medicare beneficiaries.
- To assess important outcomes, such as long-term risks and benefits, quality of life, costs, and other real-world outcomes that clinical studies have not addressed.
- To assess the comparative effectiveness of new items and services compared with existing alternatives, if not already addressed in clinical studies.
- To determine the clinical significance of statistically significant benefits documented in other studies.

But this draft document did not specify who will be required to collect or analyze the data needed for CED, or what funds will be used to support such efforts, probably because that will vary according to specific situations. For example, CED was used for coverage of off-label, unlisted uses of four drugs approved for colorectal cancer. CMS would cover such uses of these drugs only if patients enrolled in one of nine NCI-sponsored clinical trials. In this situation, NCI was responsible for gathering and analyzing the data. In contrast, CMS's CED for implantable cardiodefibrillators for primary prevention of sudden cardiac death required the implanting physician to collect the data and enter them into an existing electronic data submission system present in all hospitals. The draft guidance notes that

> Existing data systems should be used when available to avoid expending resources on creating new data systems. In addition, wherever possible, efforts should be made to use existing health information technology to support implementation of these studies. In many cases, it will be possible to link administrative data to data gathered for registries and practical trials, significantly expanding the value of the aggregate information collected and reducing the burden of data collection.

CMS will rely on the data collected to determine whether a given intervention is "reasonable and necessary for each patient who is the recipient of the item or service," a fact sheet on the guidance states (CMS, 2005b). Once collected, both CMS and the public can use the data for research purposes.

CED will be applied only in the context of a national coverage determination. However, about 90 percent of Medicare's coverage decisions are left to local carrier discretion; thus the agency expects that CED will be used infrequently (CMS, 2005b; for a review of local versus national coverage decisions, see IOM, 2001). CMS bases its legal authority for CED on its congressional mandate to provide payment only for items and services that are "reasonable and necessary" for the treatment of illness or injury. The agency claims that it needs to require CED when there is insufficient evidence to determine if a given intervention is both reasonable and necessary (CMS, 2005a, 2006).

Private insurers, however, are required to administer benefits according to the terms of the benefit plan. These plans typically exclude items or services that are deemed experimental and investigational, and they include no provision for the coverage of promising experimental interventions as evidence is developed. Some insurers have limited provision for coverage of promising experimental treatments for certain conditions (e.g., cancer, terminal illnesses) in clinical trials that meet certain qualifications. Although insurers may design benefit plans that provide CED, they may have difficulty justifying CED to plan sponsors and members. Some maintain that development of evidence of the effectiveness of an intervention is a public good that is not appropriate to the mission of private insurers, and others may question the affordability of benefit plans that provide CED (IOM, 2006). Although many insurers provide coverage of routine care costs of persons in clinical trials, they do not consider the cost of the experimental intervention itself or protocol-induced costs (costs of data collection and analysis solely for purposes of the clinical trial) as established, medically necessary treatment of the member's disease. On the other hand, some contend that investing in CED can ultimately provide payoffs to health plans both in terms of cost savings stemming from reduced use of unnecessary technologies and better outcomes for patients. Whether the knowledge gained from CED studies is proprietary or should be a public good (all health plan members, providers, and the public can benefit from the knowledge) is an unresolved issue.

The committee recommends that CMS and other health care payors, including private insurers, develop criteria for temporary, conditional coverage, similar to the CED approach, of new biomarker tests in certain circumstances to facilitate controlled and limited use of a diagnostic with a therapeutic, and even more importantly, a screening biomarker test, until sufficient evidence can be gathered to make an informed decision about standard (permanent, nonprovisional) coverage. That is, a risk-sharing approach should be implemented in which payors would agree to preliminarily cover new tests in specified circumstances contingent on data collection, to assess the clinical utility and value of the test. This would mimic the cost and risk sharing of evidence development that occurs between technology sponsors and several national health care plans overseas. Such cost and risk sharing has enabled these plans to have high standards for evidence in their coverage decisions. For example, the United Kingdom's National Health Service pays for a new drug at an agreed-on price, with the requirement that data on the drug's effectiveness be collected in a patient registry. If the drug does not show effectiveness at the expected level, the drug's price is reduced so that the total reimbursement over time reflects the actual quality of life gain observed (UK Department of Health, 2002). As noted above, private insurers may experience difficulty with CED based on their mandate and legal limitations, but it would be beneficial to examine and overcome these challenges. Because Medicare primarily covers patients who are older than 65, private insurers could make a very important contribution by collecting data on younger patient populations for whom cancer screening tests may yield the greatest gains in survival and reduced morbidity.

COST-EFFECTIVENESS ANALYSIS

The committee also recommends that when conditional coverage is applied, the cost effectiveness of biomarkers should be studied by independent research entities, in conjunction with the assessment of technology accuracy and clinical effectiveness. An independent, publicly funded information infrastructure to study and disseminate results on pharmaceutical cost-effectiveness has similarly been proposed recently (Reinhardt, 2004). Cost-effectiveness analysis (CEA) provides a framework for comparing the economic efficiencies of health care interventions. CEA measures the ratio of cost per quality-adjusted life years. This is particularly important for screening biomarkers due to the costs and potential morbidity of false-positive results (reviewed by IOM, 2001).

Although CEA is generally not used explicitly in making coverage decisions in the United States and CMS is prohibited from using CEA in making coverage decisions, there is increasing demand for cost-effectiveness analyses of medical interventions because of the rapidly increasing costs of medical care. Some experts call cost-effectiveness the "fourth hurdle" in health care, after safety, efficacy, and quality, and in some countries it is used explicitly to make coverage decisions (IOM, 2006). CEA is becoming more relevant to health policy makers because of the increasing number of options for medical interventions combined with limited financial resources and the high costs of many new medical technologies and treatments, such as the new targeted therapies for cancers.

For example, the United Kingdom's National Institute for Health and Clinical Excellence (NICE) recently decided against making two new targeted therapies, bevacizumab and cetuximab, available for the treatment of colorectal cancer on the National Health Service (NHS), arguing that neither drug is cost-effective. NICE reported that use of the drugs was not "compatible with the best use of NHS resources" because, although the treatments may extend life expectancy of some patients with advanced colorectal cancer by a few months, the average cost of treating a patient with the drugs was more than the NICE threshold of effectiveness of about £30,000 ($56,500) per life-year saved (NICE, 2006a). This figure is not an absolute ceiling, however. Taking into account the nature of the disease and quality of life provided by the drug, particularly the frequency and duration of remissions, NICE last year approved use of the drug imatinib, which targets certain types of leukemia and gastrointestinal tumors and can cost at much as £35,000 ($66,000) per year (NICE, 2006b).

Biomarker tests present another health care expense that could be a cost challenge for insurers. But their additional cost might be offset by the opportunity to better direct appropriate treatment and derive greater patient benefit for each health care dollar spent. Most cancer therapies benefit only a fraction of the patients for which they may be indicated (Spear et al., 2001). Appropriate patient selection via accurate diagnostic biomarker tests that predict responsiveness could substantially improve patient outcomes and thus increase the cost-effectiveness of treatment. Similarly, if biomarker-based screening tests could be developed to detect cancers at an earlier, more easily treated stage, these new biomarker technologies could have a substantial impact on the economic burden of cancer by reducing the cost of treatment, as well as the overall burden and consequence of disease.

But assessing the value of a biomarker diagnostic or screening test is difficult, given that such tests are intermediate steps in the patient care pathway. Because they usually trigger a cascade of decisions regarding further testing, prevention, or treatment, medical tests can have enormous influence on the ultimate costs and benefits of medical therapies. Although diagnostics account for only 1.6 percent of total Medicare costs, they influence 60 to 70 percent of downstream treatment decisions, one study found (The Lewin Group, Inc., 2005). Analytical modeling techniques may be necessary to evaluate the cost-effectiveness of new tests. Such techniques are frequently used by countries with government-funded medicine to determine how best to prioritize the health care services they provide (IOM, 2006). Modeling methods rely on the information available about the biology of disease and the effectiveness of possible interventions (IOM, 2005a). Studying the cost-effectiveness of new biomarker tests in the context of conditional coverage would facilitate methods development and help to ensure that CEA is done appropriately in the future.

Cost-effectiveness analyses assess the value of a medical treatment by noting its costs relative to its health benefits. In that way, one can choose an intervention for which the cost relative to the benefit is less than a threshold value. Health benefits are measured with an index called QALY for quality-adjusted life years. This index combines measures of quality of life with length of life. The cost-effectiveness threshold for medical interventions in the United States is between $50,000 and $100,000 per QALY (Meltzer, 2006). A cost-effectiveness analysis can be done from multiple perspectives (societal, patient, insurers, government, providers). The societal perspective is always preferred, but the committee also recommends analyzing cost-effectiveness from the insurer perspective because if the insurer and societal perspective are in conflict (e.g., an intervention is deemed cost-effective from societal perspective but not cost effective from insurer perspective), there may be a role for policy makers to intervene so that the incentives align better.

Cost-effectiveness appraisals have many methodological limitations that can affect their accuracy, as several speakers at the IOM workshop on biomarkers pointed out (IOM, 2006). How valid they are depends in part on how accurately health outcomes and other relevant metrics can be measured. The analyses can be adversely affected by basing them on inadequately controlled studies, studies that don't consider the most useful comparators, or studies that are not of sufficient duration to truly assess the health outcome of interest. The use of surrogate markers that do not adequately correlate with relevant health outcomes can also be a problem. In addition, quality-of-life measures can vary according to subpopulation, and cost assessments may not be sufficiently comprehensive. Despite their limitations, cost-effectiveness analyses are increasingly being used in the biomedical arena.

For example, to distinguish a treatable subgroup of brain cancer patients in the United Kingdom, NICE conducted a cost-effectiveness analysis to evaluate the use of cancer biomarker O6-methylguanine-DNA methyltransferase (MGMT) methylation status[2] in glioma patients. Treatment with temozolomide, in addition to radiotherapy, surpasses NICE's cost-effectiveness threshold only in the subgroup likely to respond, as indicated by MGMT methylation status. MGMT methylation status and other response-predicting biomarkers thus have the potential to refine disease and therapy and improve cost-effectiveness (Stevens, 2006).

The value of a diagnostic test, including a biomarker test, depends on how it is used. The cost-effectiveness of the Pap test substantially falls, for example, when it is used annually as opposed to every two or three years, because costs rise incrementally but benefits (years of life saved) rapidly plateau as screening frequencies increase (Eddy, 1990). Statistical analyses also reveal that self-selection of a medical treatment by patients occurs because they tend to opt out of a therapy when it is not effective. This self-selection can substantially improve the cost-effectiveness of the treatment (Meltzer, 2005). But most cost-effective analyses consider only the costs and benefits

---

[2]MGMT is a DNA-repair enzyme; its methylation inactivates the enzyme and makes it unable to repair the DNA in tumors damaged by therapy.

of a diagnostic or treatment for the entire general population and do not consider self-selection.

The cost-effectiveness of medical tests or treatments also can substantially drop if they are used incorrectly in the wrong populations. An example of this is the use of COX-2 inhibitors. Prior to the release of data showing their cardiovascular side effects, COX-2 inhibitors were shown to be relatively cost-effective drugs for patients at high risk of gastrointestinal bleeding. But the drugs were not cost-effective in people at low risk of such bleeding. However, most COX-2 inhibitors were used in the United States by people at low risk of bleeding, so the actual cost-effectiveness was poor because of how they were used (Meltzer, 2006). The cost-effectiveness of tests and interventions consequently needs to be evaluated not as they would be used under ideal circumstances, but as they are used in practice.

REIMBURSEMENT

In addition to making coverage decisions, health insurers also need to set reimbursement rates for diagnostic tests that adequately reflect their value so that they are appropriately adopted in clinical settings. On one hand, when pricing is set too low, it discourages manufacturers from developing new and innovative tests. On the other hand, generous pricing encourages rapid uptake of the test, even if widespread clinical adoption may not be justified on the basis of the evidence of a test's clinical validity or utility. When pricing is set too high, in contrast, it can impede the clinical adoption of a test. As noted in a recent Institute of Medicine (IOM) report on Medicare laboratory payment policy, "Theoretically, when prices do not reflect costs, they have the potential to inappropriately influence clinical decision making, inhibit innovation, waste taxpayer dollars, and limit beneficiary access to care" (IOM, 2000).

Medicare payment determinations for diagnostics not only affect the clinical care of its beneficiaries (one in seven patients in America) (Raab and Logue, 2001), but also influence state Medicaid and private insurers' payment rates (IOM, 2000). However, many experts argue that the reimbursement levels for diagnostics set by Medicare do not adequately reflect their cost and clinical value, with some reimbursement rates set too high relative to value, while others are too low (reviewed by The Lewin Group, Inc., 2005; IOM, 2000). The IOM assessment on Medicare's payment policy concluded (IOM, 2000):

Existing mechanisms for keeping payments up to date are inadequate . . . The process for integrating new technologies into the payment system, including determinations of coverage, assignment of billing codes, and development of appropriate prices, is slow, administratively inefficient, and closed to stakeholder participation. . . . Payments for some individual tests likely do not reflect the cost of providing services, and anticipated advances in laboratory technology will exacerbate the flaws in the current system. Problems with the outdated payment system could threaten beneficiary access to care and the use of enhanced testing methodologies in the future (pp. 7, 17).

These criticisms of Medicare's payment policy are best understood in the historical context of how Medicare determines reimbursement rates for diagnostics. When Congress enacted a Medicare clinical laboratory fee schedule in 1984, it instituted rules that served to set these reimbursement rates below market value (SSA, 1984). For example, it set the reimbursement rates offered by each of its state-wide carriers for diagnostic tests to only 60 percent of the laboratory charge current at the time. It also specified that this limit be increased each year by the consumer price index (CPI), whose rate of growth is below the rate of medical inflation. Additional legislation created national limitation amounts (NLAs), which put a cap on payment for laboratory fees that is 74 percent of the median charges and froze pricing to 1997 levels until 2009 (ignoring CPI-indicated increases) as a means of balancing the budget (Raab and Logue, 2001; AdvaMed, 2006b).

Because congressional guidance was lacking on how to determine the reimbursement rate for new tests, Medicare developed its own administrative techniques for this, without public participation. One technique, called cross-walking, creates a reimbursement rate for new tests based on how clinically or technologically similar they are to older tests with established reimbursement rates. For example, its price determination for the new iron stain test for peripheral blood is equal to its price determination for the older iron stain test for bone marrow smears (Raab and Logue, 2001). The other technique is called gap filling, for which each state carrier uses its own rules to determine the appropriate price for new tests that cannot be cross-walked. These local carrier rates are used by Medicare to determine a NLA for the new test. Local carrier payment rates that are greater than the NLA are lowered to the NLA level. But those carrier rates that are less than the NLA are not raised to the higher NLA level (Raab and Logue, 2001).

Critics cite problems with both techniques used to determine new price determinations. Both cross-walking and gap filling are inherently subjective and dependent on the technical expertise of CMS staff, which often lacks the ability to adequately judge the similarity of a new test to a test with pricing already established or to set a new fair price, some experts claim (Raab and Logue, 2001). This is compounded by another inherent problem, which is that cross-walking and gap filling are done internally, without consultation with outside experts or industry and without public commentary to correct any perceived arbitrariness or inaccuracy of final reimbursement rate determinations. Medicare and most other health plans lack test evaluation groups similar to the pharmacy and therapeutics committees they maintain to evaluate drugs (Ramsey et al., 2006). These committees are comprised of quasi-independent experts who often are not health plan employees (Ramsey et al., 2006).

Another problem inherent in Medicare's reimbursement system for laboratory tests is a coding process for such tests that is not sufficiently specific. For payors to more adequately influence the adoption of biomarker tests, those tests need to have their own Current Procedural Terminology (CPT) codes. These identifying codes are used to report medical procedures and services to health insurers and reimbursement rates are specified for each code. CPT codes are also used for developing guidelines for medical care review. Many biomarker tests do not have specific CPT codes but instead are defined by process steps, so that insurers, even if they are willing to scrutinize the clinical utility of biomarkers, often find it difficult to know what type of biomarkers are being used (IOM, 2006). This process enables biomarkers to be incorporated into clinical practice without much scrutiny.

This is especially true for homebrew tests, which are always defined by process steps. To be reimbursed, laboratories breakdown a homebrew test into specific methods and analytes used, each with its own CPT code. A single test could entail 10 to 15 different existing codes, making it difficult for the payor to discern exactly what is being tested, and eliminating the risk of seeking a new CPT code and reimbursement rate for the test. Homebrew tests thereby bypass scrutiny by both regulators and reimbursers (IOM, 2006). Even when a test has been approved by the FDA, there is no guarantee that laboratories will use that test. Instead, they may offer their own homebrew version of the test, which may not be as accurate (IOM, 2006). Homebrew versions of the HerceptTest help explain the high degree of variability in accuracy between laboratories. For example, studies suggest that the false-positive rate for the HerceptTest is as much as 48 percent greater in small laboratories that use their own homebrew versions compared with large centralized reference laboratories that use the FDA-approved version of the test (Paik et al., 2002; Perez et al., 2006; Reddy et al., 2006).

In general, there is a lack of a standardized format for the information that insurers should consider when determining what diagnostic tests to code and reimburse and what the reimbursement rate will be for those tests. This is in contrast to the format developed by the Academy of Managed Care Pharmacy (AMCP) for the evidence-based evaluation of drugs (AMCP, 2005). This format specifies the types of information that insurers should request from industry about the drugs they manufacture when making policy determinations. This information includes the drug's effectiveness and safety, its economic value relative to alternative treatments, and data on off-label indications. The AMCP format has been adopted by more than 50 health plans, hospitals, pharmacy benefit management programs, Medicaid programs, and other public agencies (Neumann, 2004). Because these guidelines are new, and because it is hard to define and measure their impact, data on the guidelines' actual impact on patient outcomes is lacking (Neuman, 2004). However, AHRQ has provided some funding to evaluate the impact of the guidelines, focusing primarily on process issues (i.e., quality of submitted dossiers).[3]

A similar format for diagnostic evaluation would offer diagnostic companies an opportunity to participate in payor decision making, providing structure and transparency for the flow of information between these two entities regarding new laboratory tests. Researchers at the Fred Hutchinson Cancer Research Center and the University of Washington in Seattle created a template for manufacturers' reporting clinical and economic information about laboratory tests that is based on the AMCP format (Ramsey et al., 2006). Other standards for evaluating diagnostic tests have been also published (Fryback and Thornbury, 1991; Reid et al., 1995). Test manufacturers or providers and health insurers could all benefit from standardizing the way evidence about new diagnostics is presented to payors (Ramsey et al., 2006). This evidence could be presented to insurers' standing committees for evaluating diagnostics akin to the pharmacy and therapeutics committees, or the test evaluation process could become an additional responsibility of already established pharmacy and therapeutics committees with the addition of appropriate expertise.

The committee recommends that CMS modernize the process for evaluating, coding, and pricing diagnostic tests. Reimbursement policies should be clarified and the decision-making process should be made more uniform and transparent. CMS should convene stakeholders to develop consensus

---

[3]See http://www.ahrq.gov/rice/ceproj.htm#Evaluation. Accessed November, 2006.

guidance on how to assess diagnostics to make coverage/reimbursement decisions. As previously recommended by an IOM report (IOM, 2000), Medicare ought to have "a single, national, rational fee schedule" for clinical laboratory tests that is based on the review of the tests by expert panels.

Similar reforms are called for in the Advanced Laboratory Diagnostics Act of 2006. This bill aims to improve the current process for determining reimbursement levels for new clinical diagnostics, correct historic payment determinations, and provide more transparency and opportunities for dialogue regarding Medicare reimbursement decisions. The Medicare Prescription Drug, Improvement, and Modernization Act, enacted in 2003, also calls for some yet-to-be implemented improvements in the coding and payment processes for new tests (AdvaMed, 2006a).

The committee also recommends that CMS use the power of its longitudinal data to assess the value of tests. Although CMS is prohibited from using clinical value as a criterion for reimbursement, assessing the clinical value of tests would aid clinical decision making.

SUMMARY AND CONCLUSIONS

Major impediments to achieving personalized medicine by implementing innovative biomarker-based cancer diagnostics in clinical settings are a lack of information about their clinical validity and utility, the inability of many diagnostic companies to expend the major resources necessary to provide this information, and inappropriate reimbursement for such diagnostics by health care payors because of an antiquated system for setting reimbursement rates.

To overcome these impediments, the committee recommends that insurers develop criteria for conditional coverage of new biomarker tests in certain circumstances, in order to allow controlled use of the tests while collecting additional information to inform final coverage decisions. The approach to conditional coverage should include the development of methods for high-quality population-based assessments of the efficacy and cost-effectiveness of biomarker tests. In addition, the committee recommends that the CMS coding and pricing system for diagnostic tests be modernized so that it more adequately fosters the appropriate reimbursement for and use of diagnostic tests.

REFERENCES

AdvaMed. 2006a. *AdvaMed Hails Bipartisan Legislation to Ensure Medicare Patient Access to New Advanced Diagnostic Laboratory Tests.* [Online]. Available: http://www.advamed. org/publicdocs/PR-332.htm [accessed September 2006].

——. 2006b. *Clinical Laboratory Diagnostic Tests Policy Milestones, 1984-2006.* [Online]. Available: http://www.advamed.org/publicdocs/clinical_lab_test_1984-2006.pdf [accessed June 5, 2006].

AMCP (Academy of Managed Care Pharmacy). 2005. *The AMCP Format for Forumlary Submissions.*

Carlos RC, Underwood W 3rd, Fendrick AM, Bernstein SJ. 2005. Behavioral associations between prostate and colon cancer screening. *Journal of the American College of Surgeons* 200(2):216-223.

CDC (Centers for Disease Control and Prevention). 2006. *Risk of Mortality from Prostate Cancer Among Men in a Randomized Trial.* [Online]. Available: http://www.cdc.gov/ cancer/prostate/screening/slides/slide20.htm [accessed July 10, 2006].

CMS (Centers for Medicare & Medicaid Services). 2005a. Draft guidance for the public, industry, and CMS staff. *Factors CMS Considers in Making Determination of Coverage with Evidence Development.*

——. 2005b. Fact Sheet: CMS responds to stakeholder feedback regarding coverage with evidence development.

——. 2006. *Guidance for the Public, Industry, and CMS Staff, National Coverage Determinations with Data Collection as a Condition of Coverage: Coverage with Evidence Development.* [Online]. Available: http://www.cms.hhs.gov/mcd/ncpc_view_document.asp?id=8 [accessed September 27, 2006].

Eddy D. 1990. Screening for cervical cancer. *Annals of Internal Medicine* 113(3):214-226.

Eifel P, Axelson JA, Costa J, Crowley J, Curran WJ Jr, Deshler A, Fulton S, Hendricks CB, Kemeny M, Kornblith AB, Louis TA, Markman M, Mayer R, Roter D. 2001. National Institutes of Health Consensus Development Conference Statement: Adjuvant therapy for breast cancer, November 1–3, 2000. *Journal of the National Cancer Institute* 93(13):979-989.

European Organisation for Research and Treatment of Cancer. July 7, 2005. *Microarray In Node Negative Disease may Avoid ChemoTherapy* (MINDACT).

Fan C, Oh DS, Wessels L, Weigelt B, Nuyten DS, Nobel AB, van't Veer LJ, Perou CM. 2006. Concordance among gene-expression-based predictors for breast cancer. *New England Journal of Medicine* 355(6):560-569.

FDA (Food and Drug Administration). 2001. Agency information collection activities; submission for OMB review; comment request; medical devices; classification/ reclassification; restricted devices: Analyte specific reagents. Notice. *Federal Register* 66:1140-1141.

Feigal E. 2006. *Partnerships to Accelerate Innovation.* Presentation at the meeting of the National Cancer Policy Forum. Washington, DC.

Feinstein AR. 2002. Misguided efforts and future challenges for research on "diagnostic tests." *Journal of Epidemiology and Community Health* 56(5):330-332.

Fisher B, Jeong JH, Bryant J, Anderson S, Dignam J, Fisher ER, Wolmark N. 2004. Treatment of lymph-node-negative, oestrogen-receptor-positive breast cancer: Long-term findings from National Surgical Adjuvant Breast and Bowel Project randomised clinical trials. *Lancet* 364(9437):858-868.

Frantz S. 2005. An array of problems. *Nature Reviews Drug Discovery* 4:362–363.

Fryback DG, Thornbury JR. 1991. The efficacy of diagnostic imaging. *Medical Decision Making* 11(2):88-94.

Goldhirsch A, Glick JH, Gelber RD, Coates AS, Senn HJ. 2001. Meeting highlights: International Consensus Panel on the Treatment of Primary Breast Cancer. Seventh International Conference on Adjuvant Therapy of Primary Breast Cancer. *Journal of Clinical Oncology* 19(18):3817-3827.

Habel LA, Shak S, Jacobs MK, Capra A, Alexander C, Pho M, Baker J, Walker M, Watson D, Hackett J, Blick NT, Greenberg D, Fehrenbacher L, Langholz B, Quesenberry CP. 2006. A population-based study of tumor gene expression and risk of breast cancer death among lymph node-negative patients. *Breast Cancer Research* 8(3):R25.

IOM (Institute of Medicine). 2000. *Medicare Laboratory Payment Policy: Now and in the Future.* Wolman DM, Kalfoglou AL, LeRoy L, eds. Washington, DC: National Academy Press.

——. 2001. *Mammography and Beyond: Developing Technologies for the Early Detection of Breast Cancer.* Nass SJ, Henderson IC, Lashof JC, eds. Washington, DC: National Academy Press.

——. 2005a. *Economic Models of Colorectal Cancer Screening in Average-Risk Adults: Workshop Summary.* Pignone M, Russell L, Wagner J, eds. Washington DC: The National Academies Press.

——. 2005b. *Saving Women's Lives.* Joy JE, Penhoet EE, Petitti DB, eds. Washington, DC: The National Academies Press.

——. 2006. *Developing Biomarker-Based Tools for Cancer Screening, Diagnosis, and Treatment: The State of the Science, Evaluation, Implementation, and Economics. A Workshop.* Patlak M, Nass S, rapporteurs. Washington DC: The National Academies Press.

The Lewin Group, Inc. July 2005. *The Value of Diagnostics Innovation, Adoption and Diffusion into Health Care.* AdvaMed. [Online]. Available: http://www.advamed.org/publicdocs/thevalueofdiagnostics.pdf. [accessed July 2006]

Meltzer D. 2005. Effects of patient self-selection on cost-effectiveness: Implications for intensive therapy for diabetes. *Society for Medical Decision Making.*

——. 2006. Cost effectiveness analysis and the value of research. Presentation at the IOM workshop on Developing Biomarker-based Tools for Cancer Screening, Diagnosis, and Treatment: The State of the Science, Evaluation, Implementation, and Economics. Washington, DC.

NCI (National Cancer Institute). 2006. Personalized treatment trial for breast cancer launched. Washington DC.

Neumann PJ. 2004. Evidence-based and value-based formulary guidelines. *Health Affairs* 23(1):124-134.

NICE (U.K. National Institute for Clinical Excellence). 2006a. *Final Appraisal Determination: Bevacizumab and Cetuximab for Metastatic Colorectal Cancer.* [Online]. Available: http://pharmalive.com/news/download.cfm?articleid=366733&attachmentid=54419 [accessed August 2006].

———. 2006b. *National Institute for Health and Clinical Excellence.* [Online]. Available: http://www.nice.org.uk/page.aspx?o=TA70; http://www.nice.org.uk/page.aspx?o=TA86 [accessed August 2006].

Paik S, Bryant J, Tan-Chiu E, Romond E, Hiller W, Park K, Brown A, Yothers G, Anderson S, Smith R, Wickerham DL, Wolmark N. 2002. Real-world performance of HER2 testing—National Surgical Adjuvant Breast and Bowel Project experience. *Journal of the National Cancer Institute* 94(11):852-854.

Paik S, Shak S, Tang G, Kim C, Baker J, Cronin M, Baehner FL, Walker MG, Watson D, Park T, Hiller W, Fisher ER, Wickerham DL, Bryant J, Wolmark N. 2004. A multigene assay to predict recurrence of tamoxifen-treated, node-negative breast cancer. *New England Journal of Medicine* 351(27):2817-2826.

Paik S, Tang G, Shak S, Kim C, Baker J, Kim W, Cronin M, Baehner FL, Watson D, Bryant J, Costantino JP, Geyer CE Jr, Wickerham DL, Wolmark N. 2006. Gene expression and benefit of chemotherapy in women with node-negative, estrogen receptor-positive breast cancer. *Journal of Clinical Oncology* 24(23):3726-3734.

Perez EA, Suman VJ, Davidson NE, Martino S, Kaufman PA, Lingle WL, Flynn PJ, Ingle JN, Visscher D, Jenkins RB. 2006. HER2 testing by local, central, and reference laboratories in specimens from the North Central Cancer Treatment Group N9831 intergroup adjuvant trial. *Journal of Clinical Oncology* 24(19):3032-3038.

Raab GG, Logue LJ. 2001. Medicare coverage of new clinical diagnostic laboratory tests: The need for coding and payment reforms. *Clinical Leadership & Management Review* 15(6):376-387.

Ramsey S, Veenstra D, Garrison L, Carlson R, Billings P, Carlson J, Sullivan S. 2006. Toward evidence-based assessment for coverage and reimbursement of laboratory-based diagnostic and genetic tests. *The American Journal of Managed Care* 12(4):21-27.

Reddy JC, Reimann JD, Anderson SM, Klein PM. 2006. Concordance between central and local laboratory HER2 testing from a community-based clinical study. *Clinical Breast Cancer* 7(2):153-157.

Reid MC, Lachs MS, Feinstein AR. 1995. Use of methodological standards in diagnostic test research. Getting better but still not good. *Journal of the American Medical Association* 274(8):645-651.

Reinhardt UE. 2004. An information infrastructure for the pharmaceutical market. *Health Affairs* 23(1):107-112.

Research Triangle Institute. 2002. *Guide to Clinical Preventive Services, Evidence Syntheses.* 3rd edition. AHRQ (Agency for Health care Research and Quality) (16):3-8.

Rogowski W. 2007. Current impact of gene technology on health care: A map of economic assessments. *Health Policy* 80(2):340-357.

Spear BB, Heath-Chiozzi M, Huff J. 2001. Clinical application of pharmacogenetics. *Trends in Molecular Medicine* 7:201-204.

SSA (Social Security Administration). 1984. Deficit Reduction Act of 1984: Provisions related to the Medicare and Medicaid programs. *Social Security Bulletin* 47(11):11-25.

Stevens, A. 2006. Cost effectiveness analysis and technology adoption in the United Kingdom. Presentation at the IOM workshop on Developing Biomarker-based Tools for Cancer Screening, Diagnosis, and Treatment: The State of the Science, Evaluation, Implementation, and Economics. Washington, DC.

Swan J, Breen N, Coates RJ, Rimer BK, Lee NC. 2003. Progress in cancer screening practices in the United States: Results from the 2000 National Health Interview Survey. *Cancer* 97(6):1528-1540.

Takahashi H, Echizen H. 2003. Pharmacogenetics of CYP2C9 and interindividual variability in anticoagulant response to warfarin. *The Pharmacogenomics Journal* 3(4):202-214.

UK Department of Health. 2002. *Drug Treatment for Multiple Sclerosis (Beta-Interferons and Glatiramer Acetate)—Risk Sharing Scheme.* [Online]. Available: http://www. dh.gov.uk/PolicyAndGuidance/OrganisationPolicy/PrimaryCare/PrimaryCareTrusts/ PrimaryCareTrustsArticle/fs/en?CONTENT_ID=4000556&chk=nvlrjJ [accessed September 2006].

van de Vijver MJ, He YD, van't Veer LJ, Dai H, Hart AA, Voskuil DW, Schreiber GJ, Peterse JL, Roberts C, Marton MJ, Parrish M, Atsma D, Witteveen A, Glas A, Delahaye L, van der Velde T, Bartelink H, Rodenhuis S, Rutgers ET, Friend SH, Bernards R. 2002. A gene-expression signature as a predictor of survival in breast cancer. *New England Journal of Medicine* 347(25):1999-2009.

Weinstein S, Obuchowski NA, Lieber ML. 2005. Clinical evaluation of diagnostic tests. *American Journal of Roentgenology* 184(1):14-19.

Acronyms and Glossary

ACRONYMS

AHRQ	Agency for Healthcare Research and Quality
ASCO	American Society of Clinical Oncology
ASR	Analyte-specific reagent
C-Path	Critical Path Institute
CDC	Centers for Disease Control and Prevention
CEA	Cost-effectiveness analysis
CED	Coverage under Evidence Development
CLIA	Clinical Laboratory Improvement Amendments
CMS	Centers for Medicare & Medicaid Services
CPI	Consumer Price Index
CPT	Current procedural terminology
CPTAC	Clinical Proteomic Technologies Initiative for Cancer
DARPA	Defense Advanced Research Projects Agency
DHHS	Department of Health and Human Services
EDRN	Early Detection Research Network
EGFR	Epidermal growth factor receptor
ER	Estrogen receptor
FDA	Food and Drug Administration
FDG-PET	fluorodeoxyglucose positron emission tomography
FNIH	Foundation of the National Institutes of Health
FTC	Federal Trade Commission

HIPAA	Health Insurance Portability and Accountability Act
IOM	Institute of Medicine
IP	Intellectual property
IRB	Institutional Review Board
IVDMIA	In Vitro Diagnostic Multivariate Index Assay
MIAME	Minimum Information About a Microarray Experiment
MMRC	Multiple Myeloma Research Consortium
NBN	National Biospecimen Network
NCI	National Cancer Institute
NICE	U.K. National Institute for Clinical Excellence
NIH	National Institutes of Health
NIST	National Institute of Standards and Technology
NLA	National limitation amount
OBQI	Oncology Biomarker Qualification Initiative
PBRC	Pharmaceutical Biomedical Research Consortium
PET	Positron emission tomography
PMA	Premarket approval
PPP	Public–private partnerships
PSA	Prostate-specific antigen
PSI	Protein Structure Initiative
SACGT	Secretary's Advisory Committee on Genetic Testing
SEMATECH	Semiconductor Manufacturing Technology
SNP	Single nucleotide polymorphism
TRWG	Translational Research Working Group
TSC	The SNP Consortium
USPSTF	U.S. Preventive Services Task Force

GLOSSARY

Allele—any one of a series of two or more different genes that occupy the same position (locus) on a chromosome.

Amplification—a process by which specific genetic material is increased. For some cancers, the number of copies of specific genes is higher than normal. These genes are said to be amplified.

Analyte-specific reagent (ASR)—antibodies, both polyclonal and monoclonal, specific receptor proteins, ligands, nucleic acid sequences, and similar reagents, which through specific binding or chemical reaction with substances in a specimen are intended to be used in a diagnostic application for identification and quantification of an individual chemical substance or ligand in biological specimens.

Analytical validity—the accuracy of a test in detecting the specific entity that it was designed to detect. This accuracy does not imply any clinical significance, such as diagnosis.

Bias—the systematic but unintentional erroneous association of some characteristic with a group in a way that distorts a comparison with another group.

Biorepository—a collection of biological samples, such as tissue, that can be used for research.

BRCA—a gene that when mutated increases a woman's risk of developing breast cancer. Two BRCA genes have been identified and are known as BRCA1 and BRCA2.

Cetuximab—a monoclonal antibody drug used to treat advanced or metastatic cancer of the colon and rectum, usually in combination with chemotherapy or irinotecan, another cancer drug. It is currently being used in research trials for treatment of head and neck cancers.

Clinical endpoint—a characteristic or variable that reflects how a patient feels, functions, or survives in response to an intervention.

Clinical trial—a formal study carried out according to a prospectively defined protocol that is intended to discover or verify the safety and effectiveness of procedures or interventions in humans.

Clinical utility—the clinical and psychological benefits and risks of positive and negative results of a given technique or test.

Clinical validity—the accuracy of a test for a specific clinical purpose, such as diagnosing or predicting risk for a disorder.

Comparative genomic hybridization—a technique for detecting the gain or loss of genetic material in tumor cells.

Computed tomography (CT)—a special radiographic technique that uses a computer to assimilate multiple X-ray images into a two-dimensional, cross-sectional image, which also can be reconstructed into a three-dimensional image. This can reveal many soft tissue structures not shown by conventional radiography.

Conditional coverage—a policy by which insurers agree to preliminarily cover new tests with the proviso that data would be collected in conjunction with the use of the test, to assess the clinical utility and value of the test, and to create better evidence. Data collected during conditional coverage assessments are used in later decisions regarding full coverage and may be used for research purposes afterward.

Coverage with evidence development (CED)—a CMS program whereby prospective data collection on a product is required for national Medicare coverage (see Conditional coverage). A product that has an insufficient evidence base for CMS coverage determination could be evaluated through CED.

Current Procedural Terminology—a listing of descriptive terms and identifying codes for reporting medical services and procedures, designed to standardize the terminology used for medical, surgical, and diagnostic services. CPT codes were first developed by the American Medical Association and are updated by the CPT Editorial Panel.

CYP450—the gene that codes for the drug-metabolizing enzyme cytochrome P450. Variants in this gene can alter the enzyme's ability to metabolize certain drugs.

Defense Advanced Research Projects Agency (DARPA)—the central research and development organization for the Department of Defense, it has focused on research projects that have high risk but also potential for high payoff if successful.

De novo classification—a Food and Drug Administration classification of a device or diagnostic that is not equivalent to a legally marketed product.

Deletion—the loss of genetic material. Some cancers are triggered by the deletion of key genes, portions of genes, or their regulatory sequences.

Diagnostic—the investigative tools and techniques used in biological studies or to identify or determine the presence of a disease or other condition. In this report, "diagnostic" is often used synonymously with "biomarker test." These terms refer to any laboratory-based test that can be used in drug discovery and development as well as in patient care and clinical decision making.

Epidermal growth factor receptor (EGFR)—a receptor that is overproduced in several solid tumors, including breast and lung cancers. Its overproduction is linked to a poorer prognosis because it enables cell proliferation, migration, and the development of blood vessels. Several new drugs recently approved by the Food and Drug Administration specifically target EGFR.

Flow cytometry—a technique for identifying and sorting cells and their components (such as DNA) by staining with fluorescent dyes and detecting the fluorescence, usually by laser beam illumination.

Genome—an organism's entire complement of DNA, which determines its genetic makeup.

Genomics—the study of all of the nucleotide sequences, including structural genes, regulatory sequences, and noncoding DNA segments, in the chromosomes of an organism or tissue sample. One example of the application of genomics in oncology is the use of microarray or other techniques to uncover the genetic "fingerprint" of a tissue sample. This genetic fingerprint is the pattern that stems from the variable expression of different genes in normal and cancer tissues.

Genotype—the genetic makeup of an organism or cell.

Health Insurance Portability and Accountability Act (HIPAA)—an act passed in 1996 that includes privacy and security regulations regarding disclosure and use of medical information.

Herceptin—see Human epidermal growth factor receptor 2.

High-density expression arrays—microarrays with so many probes that they can detect the expression of hundreds of thousands of genes, as opposed to low-density expression arrays, which can detect a much smaller number.

High-throughput technology—any approach using robotics, automated machines, and computers to process many samples at once.

Homebrew test—diagnostic tests that are custom made in individual laboratories by combining several reagents in a specified protocol. All testing of a homebrew diagnostic is done within the laboratory that developed it. The Food and Drug Administration regulates commercial tests through a premarket approval (PMA) or premarket notification (510[k]) review process. In contrast, it does not regulate homebrew tests, except to the extent that they use analyte-specific reagents. Clearance or approval of the test itself is not required.

Human epidermal growth factor receptor 2 (HER2)—a growth factor receptor that is used as a breast cancer biomarker for prognosis and treatment with the drug trastuzumab (Herceptin), which targets the protein. The HER2 protein is overexpressed in approximately 25 percent of breast cancer patients, due to amplification of the gene.

Human Genome Project—a 13-year project coordinated by the U.S. Department of Energy and the National Institutes of Health and completed in 2003. The project completed its goal of sequencing the genome and mapping all 20,000–25,000 genes in human DNA two years earlier than anticipated, due to technological advances.

Imatinib—A small molecule compound originally developed for treating chronic myelogenous leukemia and gastrointestinal stromal tumors, imatinib (STI571, Gleevec) is a selective tyrosine kinase inhibitor that binds to the ATP-binding pocket and blocks the tyrosine kinase activities of Abl, c-kit, and PDGFR.

Liquid chromatography—a process in which a chemical mixture carried by a liquid is separated into its components due to the different rates at which these components travel through a stationary phase.

Loss of heterozygosity—loss of one allele at a specific position on a chromosome.

Magnetic resonance imaging (MRI)—method by which images are created by recording signals generated from the excitation (the gain and loss of energy) of such elements as the hydrogen of water in tissue when placed in a powerful magnetic field and pulsed with radio frequencies.

Mass spectrometry—a method for separating ionized molecular particles according to mass by applying a combination of electrical and magnetic fields to deflect ions passing in a beam through the instrument.

Messenger RNA (mRNA) expression profiling—the use of microarrays or other technology to quantify all the different mRNAs transcribed from the various protein-encoding genes in a sample. (Messenger RNA carries the information from the DNA genetic code to areas in the cytoplasm of the cell in which proteins are made.)

Metabolomics—the systematic study of the unique chemical fingerprints that specific cellular processes leave behind, that is, small-molecule metabolites.

Microarray—a high-throughput biological assay in which different probes are deposited on a chip surface (glass or silicon) in a miniature arrangement. DNA microarrays are the most commonly used.

Off-label use—using a drug that either has not been approved by the Food and Drug Administration or has not been approved for the purpose for which it is being used.

Overfitting—a false pattern that is found between large numbers of possible predictors and an outcome due to high complexity and "noise" in the data. Overfitting leads to erroneous conclusions about the data. This can be identified by checking the reproducibility in a separate, independent group of individuals.

Pathway biomarker—a biomarker that can be detected in one or several key steps along a biochemical pathway that may be perturbed in cancer cells. Because of their broad applicability, pathway biomarkers may be useful in assessing the effectiveness of multiple drugs in different types of cancers.

Pharmacodynamics—the study of the biochemical and physiological effects of drugs, the mechanisms of drug action, and the relationship between drug concentration and effect. Pharmacodynamics is the study of what a drug does to the body, as opposed to pharmacokinetics, which is the study of what a body does to a drug.

Pharmacogenomics—a biotechnological science that combines the techniques of medicine, pharmacology, and genomics to determine the effects of genetic differences in patients on the metabolism and hence the potential toxicity or efficacy of drugs.

Pharmacokinetics—the study of the time course of substances, such as drugs, in an organism. Pharmacokinetics is used to determine how long a drug remains in the body.

Phase I trial—clinical trial in a small number of patients in which the toxicity and dosing of an intervention are assessed.

Phase II trial—clinical trial in which the safety and preliminary efficacy of an intervention are assessed in patients.

Phase III trial—large-scale clinical trial in which the safety and efficacy of an intervention are assessed in a large number of patients. The Food and Drug Administration generally requires new drugs to be tested in phase III trials before they can be put on the market.

Phenotype—the physical traits of an individual.

Phosphorylated proteins—proteins to which a phosphate group has been attached. The excessive growth that typifies cancer is often thought to be prompted by the phosphorylation of growth-signaling proteins called tyrosine kinases. Such phosphorylation activates these enzymes which then phosphorylate other molecules.

Polyacrylamide gel electrophoresis (two-dimensional)—a technique used to separate molecules out of a solution based on their charge, isoelectric point, mass, and size. One-dimensional electrophoresis, in contrast, has fewer molecule-distinguishing capabilities, as it only separates molecules out of a solution on the basis of their charge and size.

Polymerase chain reaction (PCR)—a technique for duplicating genetic sequences in vitro by as many as a billion times. This technique enables the detection of relatively scarce genetic material.

Polymorphism—existence of a gene in several allelic forms.

Positive predictive value—the probability that an individual with a positive test has, or will develop, a particular disease, or characteristic, that the test is designed to detect. It is a measure of the ratio of true positives to (false + true positives).

Positron emission tomography (PET)—a highly sensitive technique that uses radioactive probes to image in vivo tumors, receptors, enzymes, DNA replication, gene expression, antibodies, hormones, drugs, and other compounds and processes.

Premarket approval (PMA)—a Food and Drug Administration approval for a new test or device that enables it to be marketed for clinical use. To receive this approval, the manufacturer of the product must submit clinical data showing the product is safe and effective for its intended use.

Premarket notification or 510(k)—a Food and Drug Administration review process that enables a new test or device to be marketed for clinical use. This review process requires manufacturers to submit data showing the accuracy and precision of their product, as well as, in some cases, its analytical sensitivity and specificity. Manufacturers also have to provide documentation supporting the claim that their product is substantially equivalent to one already on the market. This review does not typically consider the clinical safety and effectiveness of the product.

Proficiency testing—laboratories performing nonwaived tests must enroll laboratory personnel in tests specific to the subspecialty relevant to the tests they will be evaluating. The Clinical Laboratory Improvement Act requires proficiency testing of personnel at least once every two years.

Protein chip—a piece of glass or other surface on which different protein probes have been affixed at separate locations in an ordered manner. The probes are often antibodies to specific proteins. The protein chip identifies the amounts and types of proteins present in a sample via fluorescence-based imaging.

Proteomics—the study of the structure, function, and interactions of the proteins produced by the genes of a particular cell, tissue, or organism. The application of proteomics in oncology may involve mass spectroscopy, two-dimensional polyacrylamide gel electrophoresis, protein chips, and other techniques to uncover the protein "fingerprint" of a tissue sample. This protein fingerprint is the pattern that stems from the various amounts and types of all the proteins in the sample.

PSA test—a blood test that detects prostate-specific antigen. The PSA test was approved by the Food and Drug Administration in 1985 for prostate cancer recurrence, but it is now widely used as a screening test for prostate cancer.

Qualification—the evidentiary process of linking an assay with biological and clinical endpoints that is dependent on the intended application.

Quality-adjusted life-year (QALY) index—an index that combines measures of quality of life with length of life.

Sample bias—see Bias.

Sensitivity (analytical)—the lowest concentration that can be distinguished from background noise. This concentration is termed an assay's detection limit.

Sensitivity (clinical)—a measure of how often a test correctly identifies patients with a specific diagnosis. It is calculated as the number of true-positive results divided by the number of true-positive plus false-negative results.

Single-molecule sequencing—also called nanopore sequencing, is a method for sequencing DNA that involves passing the DNA through small pores about 1 nanometer in diameter. The size of the pore ensures that the DNA is forced through the hole as a long string, one base at a time. The base (i.e., adenine, guanine, cytosine, or thymine) is identified by the characteristic obstruction it creates in the pore, which is detected electrically. Single-molecule sequencing can be a more sensitive technique for identifying relatively rare genetic strands in a sample, without the need for replicating them with a polymerase chain reaction.

Single nucleotide polymorphism (SNP)—a variant DNA sequence in which the purine or pyrimidine base (e.g., cytosine) of a single nucleotide has been replaced by another such base (e.g., thymine).

SNP microarray—a type of microarray used to identify genetic changes linked to specific cancers.

Specificity (analytical)—how well an assay detects only a specific substance and does not detect closely related substances.

Specificity (clinical)—a measure of how often a test correctly identifies the proportion of persons without a specific diagnosis. It is calculated as the number of true-negative results divided by the number of true negative plus false-positive results.

Surface-enhanced laser desorption/ionization (SELDI)—a technique that uses chemical or antibody probes to bind to specific proteins in a sample. The bound proteins are then vaporized with a laser and ionized for analysis in a mass spectrometer. Patterns of the masses of the various proteins in a sample, rather than actual protein identifications, are produced by SELDI analysis. These mass spectral patterns are used to differentiate patient samples from one another, such as to distinguish diseased from normal samples.

Surrogate endpoint—a biomarker that is intended to substitute for a clinical endpoint in a therapeutic clinical trial and is expected to predict clinical benefit (or harm or lack of benefit or harm) based on epidemiologic, therapeutic, pathophysiologic, or other scientific evidence.

Trastuzumab—see HER2.

Validation—the process of assessing the assay or measurement performance characteristics.

Appendix

Developing Biomarker-Based Tools for Cancer Screening, Diagnosis,and Treatment: The State of the Science, Evaluation, Implementation, and Economics, Workshop Summary

Margie Patlak and Sharyl Nass, Rapporteurs

THE NATIONAL ACADEMIES PRESS   500 Fifth Street, N.W.   Washington, DC 20001

NOTICE: The project that is the subject of this report was approved by the Governing Board of the National Research Council, whose members are drawn from the councils of the National Academy of Sciences, the National Academy of Engineering, and the Institute of Medicine. The members of the committee responsible for the report were chosen for their special competences and with regard for appropriate balance.

This study was supported by Contract Nos. HHSH25056133, HHSN261200611002C, 200-2005-13434, HHSM-500-2005-00179P, HHSP23320042509XI, and 223-01-2460 between the National Academy of Sciences and the Department of Health and Human Services. Any opinions, findings, conclusions, or recommendations expressed in this publication are those of the author(s) and do not necessarily reflect the view of the organizations or agencies that provided support for this project.

International Standard Book Number-10   0-309-10134-4
International Standard Book Number-13   978-0-309-10134-9

Additional copies of this report are available from the National Academies Press, 500 Fifth Street, N.W., Lockbox 285, Washington, DC 20055; (800) 624-6242 or (202) 334-3313 (in the Washington metropolitan area); Internet, http://www.nap.edu.

For more information about the Institute of Medicine, visit the IOM home page at: **www.iom.edu.**

The serpent has been a symbol of long life, healing, and knowledge among almost all cultures and religions since the beginning of recorded history. The serpent adopted as a logotype by the Institute of Medicine is a relief carving from ancient Greece, now held by the Staatliche Museen in Berlin.

Developing Biomarker-Based Tools
for Cancer Screening,
Diagnosis, and Treatment

INTRODUCTION

Research has long sought to identify biomarkers that could detect cancer at an early stage, or predict the optimal cancer therapy for specific patients. Fueling interest in this research are recent technological advances in genomics, proteomics, and metabolomics that can enable researchers to capture the molecular fingerprints of specific cancers and fine-tune their classification according to the molecular defects they harbor. The discovery and development of new markers of cancer could potentially improve cancer screening, diagnosis, and treatment. Given the potential impact cancer biomarkers could have on the cost effectiveness of cancer detection and treatment, they could profoundly alter the economic burden of cancer as well.

Despite the promise of cancer biomarkers, few biomarker-based cancer tests have entered the market, and the translation of research findings on cancer biomarkers into clinically useful tests seems to be lagging. This is perhaps not surprising given the technical, financial, regulatory, and social challenges linked to the discovery, development, validation, and incorporation of biomarker tests into clinical practice. To explore those challenges and ways to overcome them, the National Cancer Policy Forum held the conference "Developing Biomarker-Based Tools for Cancer Screening, Diagnosis and Treatment: The State of the Science, Evaluation, Implementation, and Economics" in Washington, D.C., from March 20 to 22, 2006.

At this conference, experts gave presentations in one of six sessions:

- Brief overview of technologies, including genomics, proteomics, metabolomics, and functional imaging
- Overcoming the technical obstacles, with presentations on informatics and data standards, and biomarker validation and qualification
- Coordinating the development of biomarkers and targeted therapies, with a clinical investigator and representatives from industry and the National Cancer Institute offering their perspectives
- Biomarker development and regulatory oversight, including current regulations governing biomarker tests as well as new clinical trial designs needed to incorporate biomarker tests that predict patient responders
- Adoption of biomarker-based technologies, with discussion on what motivates private insurers and Medicare to cover biomarker-based tests and what various organizations consider when recommending such tests be adopted into clinical practice
- Economic impact of biomarker technologies, with an exploration of cost-effectiveness analyses of biomarker tests and a payor perspective on the evaluation of such tests

In addition, seven small group discussions explored the policy implications surrounding biomarker development and adoption into clinical practice:

- Clinical development strategies for biomarker utilization
- Strategies for implementing standardized biorepositories
- Strategies for determining analytic validity and clinical utility of biomarkers
- Strategies to develop biomarkers for early detection
- Mechanisms for developing an evidence base
- Evaluation of evidence in decision making
- Incorporating biomarker evidence into clinical practice

This document is a summary of the conference proceedings, which will be used by an Institute of Medicine (IOM) committee to develop consensus-based recommendations for moving the field of cancer biomarkers forward. The views expressed in this summary are those of the speakers and discussants, as attributed to them, and are not the consensus views of the participants of the workshop or of the members of National Cancer Policy Forum.

OVERVIEW OF TECHNOLOGIES USED TO DISCOVER CANCER BIOMARKERS

Technology is constantly evolving and recent technological advances have made it easier to discover many potential cancer biomarkers through high-throughput screens. Advances in imaging technology also are furthering the discovery and use of biomarkers. The goal of the first session of the conference was to provide a brief overview of the technologies currently being used to identify and develop cancer biomarkers (Figure 1).

Genomics, Proteomics, and Metabolomics

Todd Golub, MD, of the Dana-Farber Cancer Institute, began by discussing several of the genomics-based techniques commonly used to discover biomarkers for cancer detection or for patient stratification for therapy. Some of these techniques detect changes at the DNA level (are DNA-based), whereas others detect changes at the RNA level and are considered RNA-based.

Dr. Golub explored which type of genomics test—DNA based or RNA based—would be likely to serve as a better biomarker if cost were not an issue. DNA-based tests are advantageous because DNA is more stable than RNA, and because most changes related to cancer occur at the DNA level, he said. But he noted that perhaps one could make a stronger argument for RNA-based tests because not only can they detect oncogenic RNA missteps, but molecular signatures at the RNA level also help reveal upstream DNA-level abnormalities that could contribute to a cancer. These abnormalities include base substitutions, and amplifications or deletions that alter the copy number or heterozygosity of specific genetic sequences. Dr. Golub noted that studying epigenetic changes in DNA, such as methylation, and genome rearrangements, such as chromosome translocations, can also lead to discovery of important cancer biomarkers, although he did not have time to address these topics in his presentation.

Although early genetic analyses of cancers focused on detecting changes in the copy number of genes, Dr. Golub stressed that it is also important to screen for loss of heterozygosity (LOH). LOH can occur without a change

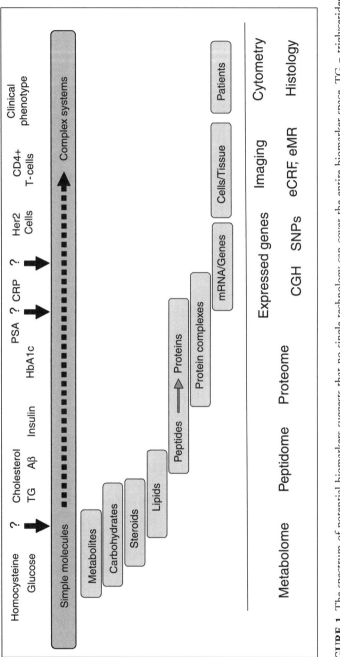

FIGURE 1 The spectrum of potential biomarkers suggests that no single technology can cover the entire biomarker space. TG = triglycerides; Aβ = β-amyloid; HbA1c = hemoglobin A1c; PSA = prostate-specific antigen; CRP = C-reactive protein; CGH = comparative genomic hybridization; SNP = single nucleotide polymorphism; eCRF = electronic case report form; eMR = electronic medical record.
SOURCE: Adapted from Schulman presentation (March 20, 2006).

in gene copy number, he noted, if both alleles for a specific gene have been mutated or epigenetically altered. This copy-neutral LOH may account for as much as half of all LOH in the genome.

Two main types of arrays are used to detect changes in copy number or LOH linked to cancer. Single nucleotide polymorphism (SNP) arrays have between 50k and 500k SNPs across the genome and can detect both copy number changes and other forms of LOH. Comparative genomic hybridization arrays can detect changes in copy number of DNA content, but are unable to detect LOH in which the copy number remains the same. For this reason, Dr. Golub prefers SNP arrays for detecting cancer biomarkers. Higher density SNP arrays can give sharper resolution by reducing the signal-to-noise ratio than lower density SNP arrays, he pointed out. But the optimal amount of density that is the most cost-efficient means for detecting cancer biomarkers remains to be determined.

Standard DNA sequence analysis of tumor samples as a means of detecting cancer biomarkers has numerous drawbacks, which Dr. Golub pointed out. Not only is it difficult and costly to do, but it is frequently inaccurate, causing false negatives because of normal tissue contamination of the tumor samples used. Most tumor samples contain a mixture of normal cells, such as inflammatory cells, as well as tumor cells. Because the Sanger sequencing results are an average of both the normal and tumor cells in a sample, normal genome contamination can obscure mutations in tumor cells that might serve as cancer biomarkers.

However, newer techniques, such as single-molecule sequencing, may substantially lower the cost of sequencing, and should avoid the problems of normal cell contamination that plague standard sequencing efforts. "I think this is exactly the type of technology, even if cost neutral, that would dramatically accelerate our ability to detect important mutations in cancer," Dr. Golub said.

To exemplify this, Dr. Golub reported on results from his colleagues at Dana-Farber who used single-molecule sequencing to detect a mutation that was linked to resistance of the drug Iressa in a lung cancer patient. The lung fluid sample the researchers analyzed only had 3 percent tumor cells, and a standard Sanger sequencing analysis missed the mutation.

Once a genetic signature with likely clinical relevance has been discovered, custom-made arrays that only have the gene sequences of interest need to be made for preclinical or clinical testing. Dr. Golub described a few genetic signature amplification and detection platforms useful for such testing, including a Luminex bead-based method. For this method, the genetic material is amplified using polymerase chain reaction. The genetic signature is then read not on microarrays, but on miniscule color-coded beads that are detected by lasers in a flow cytometer. This is an inexpensive way to detect genetic signatures, costing about 50 cents for every 100 transcript signatures. One can also use the standard mRNA expression profiling platforms that are commercially available. These are all sufficiently accurate and precise to be used in a clinical setting to detect genetic signatures, according to Dr. Golub. Cost and throughput will be significant drivers of this technology, he added.

The next presentation was on proteomics and metabolomics technologies, given by Howard Schulman, PhD, of PPD Biomarker Discovery Sciences. One of those techniques, which Dr. Schulman described as the traditional proteomic workhorse, uses two-dimensional polyacrylamide gels for the separation stage. This is a slow process that is less amenable to high-throughput. Surface-enhanced laser desorption/ionization is a high-throughput technology that can more quickly separate the proteins in a sample, but identifying the protein peaks is a challenge. That identification process can be bypassed by using software to differentially identify patterns of protein peaks to find a molecular fingerprint that can distinguish cancerous from noncancerous tissue. This fingerprint is based on the amounts of all the various proteins detected, without knowledge of what those individual proteins are, Dr. Schulman noted. However, it can be problematic to translate mass spectroscopy fingerprints into a clinical diagnostic test without identifying or further characterizing those proteins.

One- and multidimensional liquid chromatography are also used to separate peptides in a sample (after protein digestion) that a mass spectrometer can differentially quantify and then identify (Figure 2). But the amplitude for each of the peptides can vary depending on the composition of the mixture, which makes it hard to compare one person's sample with another's, and one batch run versus another. This has proven problematic for researchers trying to develop cancer biomarkers based on differential quantification, otherwise known as molecular fingerprinting.

To improve such differential accuracy, researchers developed a method called isotope-coded affinity tags several years ago. This technique labels a portion of a sample with a mass tag and runs both labeled and unlabeled samples through the mass spectrometer at the same time. The labeled sample serves as a sort of baseline control for the unlabeled sample. This helps normalize or eliminate a lot of the peak amplitude variability due to differences in mixture composition. But this is a more costly method because

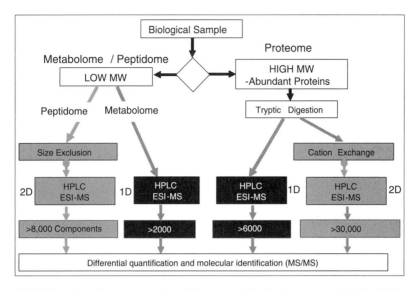

FIGURE 2 One-dimensional and multidimensional liquid chromatography LC-LC/MS. LC = liquid chromatography; MS = mass spectroscopy; MW = molecular weight; HPLC = high-performance liquid chromatography; ESI = electrospray ionization. SOURCE: Schulman presentation (March 20, 2006).

of the need for the reagents, and it has some bias introduced by the type of tag used, according to Dr. Schulman. The field is rapidly adopting a label-free approach in which chromatographic separation techniques and mass spectrometry are coupled with software-based solutions for normalizing the variation in amplitude signal due to differences in mixture composition to yield accurate differential expression data.

Dr. Schulman concluded his talk by noting that the current state of proteomics is comparable to the early days of microarrays, which could detect about one-sixth the number of genetic sequences that can now be detected. But proteomics is still highly effective even without the ability to profile every protein, he said. He noted that one can profile more than a thousand proteins by using multidimensional chromatography. But the tradeoffs with more fractionation are lower throughput (due to slower processing) and higher costs. The advantage of proteomic and metabolomic profiling is that you can sample readily accessible tissues, such as plasma and urine, that are ideal for monitoring biomarkers in clinical trials and testing diagnostics.

He also noted that the lowest abundance proteins, such as cytokines or other signaling molecules, will likely require antibody-based protein chips to complement liquid chromatography separation techniques. Sensitivity to such proteins could also be increased by using samples likely to have higher concentrations of biomarkers of interest. For example, cerebral spinal fluid could be tapped to find biomarkers for lymphoma metastases in the central nervous system, or prostatic fluid could be used to detect biomarkers for prostate cancer. Affinity capture of protein subcategories, such as phosphorylated proteins, could also selectively profile lower abundance proteins of interest.

Drs. Schulman and Golub stressed the need to experimentally validate the biological basis and importance of detected genetic or proteomic differences in a disease process. For example, researchers in Dr. Golub's laboratory used high-density expression arrays to detect an RNA signature in bone marrow samples that correlated with response to a drug for myelodysplastic syndrome. They found a group of genes that were only highly expressed in patients who responded to the drug. Many of these genes previously had been identified as markers for late red blood cell differentiation, leading to the hypothesis that such differentiation may be predictive of drug response.

To test this idea, they induced normal immature blood cells to differentiate into red blood cells. They found that all of the genes, whose boosted expression was linked to drug response in their biomarker discovery study, also had heightened expression during the red blood cell differentiation that occurred in their experiments. This validated their hypothesis and put the concept of genetic signature for drug response on firmer footing. "The most valuable and robust biomarkers will be those that have some component of experimental validation accompanying them," Dr. Golub said. He added that "the challenge looking forward is going to be to move from simply cataloging mutations or genome abnormalities to coalescing those abnormalities into more of a molecular taxonomy that brings biological understanding to this catalog. The more we can integrate these anonymous molecular signatures with biological knowledge, the more they're likely to stick."

Dr. Golub also pointed out the need to develop biomarker diagnostics that can easily be used on the paraffin-embedded or formalin-fixed tissues that are routinely collected in the clinic. "We need to make the technology work for those routinely collected samples rather than retrain the medical community to collect samples in a different way," Dr. Golub said.

Drs. Golub and Schulman noted that a lack of good-quality samples can be a stumbling block for biomarker discovery. Rarely are enough samples collected in a clinical trial, and those samples that are collected are usually fixed in formalin, which can affect their ability to be analyzed in a mass spectrometer. Dr. Schulman suggested that pharmaceutical and biotechnology companies have experimental medicine groups that are best positioned to collect the samples required to discover biomarkers.

But the biggest impediment for biomarker development, which Drs. Golub and Schulman both cited, was a lack of a critical mass of research in the discovery phase. "The bottleneck is not so much on the regulatory side or the validation side, but that not enough of the discovery effort has been made," said Dr. Schulman.

As to whether such efforts at biomarker discovery should take a hypothesis-driven or open-ended approach, Drs. Golub and Schulman agreed that both approaches were necessary. Open-ended discovery aims at uncovering a molecular understanding of a particular type of cancer that may eventually lead to useful biomarkers. A hypothesis-driven approach, in contrast, is more streamlined at finding molecular changes likely to predict a response to therapy or some other useful clinical endpoint. There is a role for both these approaches, Dr. Schulman said. But he added that pharmaceutical companies are unfortunately more likely to conduct a hypothesis-driven search for biomarkers that predict drug response than to support a more open-ended search. Dr. Golub noted that the danger of conducting only hypothesis-driven research on biomarkers is that it does not address the challenge of "how do we get beyond discovering what we already know, in terms of biological knowledge?"

Molecular Imaging

Next, Michael Phelps, PhD, of the University of California, Los Angeles, discussed molecular imaging biomarkers for drug discovery, development, and patient care. He described how positron emission tomography (PET) can be used as a molecular camera to image *in vivo* processes at the molecular level. But PET is more than an imaging device, as it also can be used analytically to perform a variety of quantitative biochemical and biological assays.

There are currently about 600 PET probes for metabolism, receptors, enzymes, DNA replication, gene expression, antibodies, hormones, drugs, and other compounds in nanomole amounts. Typical antibody probes get broken down too quickly to be effective for PET imaging, but there are modified antibody probes that are small molecule versions of the original antibodies and retain the active end. Most PET probes were developed from probes used in *in vitro* assays so as to translate that assay into an *in vivo* measurement. Ninety percent of PET probes were developed from drugs, Dr. Phelps reported.

Over the past few years, PET scanners have merged with computed tomography (CT) scanners to combine the anatomical definition of the CT with the biological assay capability of the PET scan. Researchers have also created microPET/CT machines to image biological processes and drug responses in mice.

Because PET probes are administered in nanomole amounts, measures can be performed on biological processes without disturbing the processes or causing pharmacologic mass effects, Dr. Phelps noted. Not only can PET scans be safely done, but studies show they are more accurate than magnetic resonance imaging or CT scans for the diagnosis and staging of cancer, for assessing therapeutic response, and for detecting cancer recurrence.

To detect cancer, technicians usually use a PET probe that images the heightened glucose metabolism of cancer cells. To predict or determine response to therapy, a number of different types of probes are used, depending on the type of cancer and type of treatment. The PET assay can enable stratification of patients according to whether they have the therapeutic target. For example, a probe that detects DNA replication may be used to predict whether a cancer will respond to a chemotherapy that blocks such replication. A probe for an estrogen receptor may be used to determine if breast cancer metastases are likely to respond to hormonal therapy. PET is especially useful for revealing whether a tumor is responding to therapy. It can detect within a day, for example, whether patients' tumors are not responding to Gleevec, thereby quickly determining if patients should receive a different drug, Dr. Phelps pointed out.

PET imaging also has advantages over standard techniques for assessing the pharmacokinetics and pharmacodynamics of drugs, he added. For example, standard pharmacokinetic assessments are based on measurements of how quickly a drug is cleared from the blood. In contrast, by using labeled drugs as probes, PET can precisely measure the concentration of the drug, not just in the blood, but in all tissues over time, he noted.

Dr. Phelps described a recent innovation in PET technology that uses "click chemistry" to create PET probes. This technique involves combining two small molecules with low to moderate affinity to the target, but high affinity to each other. They collectively latch onto the target as they bind to each other. The end result is that they bind to the target with an extremely high affinity that is the product of the affinities of the two molecules. These probes dramatically increase the sensitivity and physical resolution of PET imaging. Because the probes are such small molecules, they can access surface receptors, cells, and even the cell nucleus.

Dr. Phelps concluded his talk by noting there are "PET pharmacies" scattered all over the world that use automated chemistry to make and ship labeled molecular PET probes. There are also "labs on a chip" that enable researchers to custom build their own PET probes using click chemistry and other techniques.

In response to comments by Drs. Golub and Schulman regarding where the bottleneck is in biomarker development, Dr. Phelps noted that as one gets closer to introducing a biomarker into a clinical setting, Food and Drug Admininstration (FDA) premarket regulation can become very limiting. He suggested that regulatory bodies work with researchers to change the criteria by which drugs and molecular diagnostics are evaluated.

MEETING THE TECHNICAL CHALLENGES OF BIOMARKER VALIDATION AND QUALIFICATION

Appropriate analysis and interpretation of biomarker data presents enormous challenges, especially with the advent of genomic and proteomic technologies that can generate a tremendous amount of data on individual samples. Three speakers at the conference addressed the technical challenges involved with validating the accuracy and clinical relevance of cancer biomarkers. John Quackenbush, PhD, of Harvard University spoke about experimental design considerations and data reporting standards to aid the validation of biomarkers. David Ransohoff, MD, of the University of North Carolina also discussed experimental design, and the shortcomings of recent cancer biomarker studies that should be avoided in future studies. John Wagner, MD, PhD, of Merck and Co., Inc., gave a pharmaceutical company's perspective on what is required to validate a cancer biomarker and establish its relevance to useful clinical endpoints.

Dr. Quackenbush began this session by noting that with microarray technologies, researchers tend to do more hypothesis-generating experiments than hypothesis-driven experiments. But despite a lack of an experimental hypothesis, one still needs to think critically about experimental design and how data are collected, managed, and analyzed, he said. All of these steps play crucial roles in determining whether the results derived from biomarker studies are clinically meaningful and valid in broader populations than in the original test population.

Drs. Quackenbush and Ransohoff stressed that the same issues that apply to standard hypothesis-driven clinical studies are also applicable to studies in genomics, proteomics, and metabolomics, which they collectively called "omics." "This is an exciting era because we have very powerful tools to measure the biology [of cancer], but the rules of evidence about validity have not changed," said Dr. Ransohoff. "New reductionist methods mean lots more data, but not necessarily more knowledge, and the rules of evidence about how you go from data to knowledge haven't changed."

Dr. Quackenbush cited a need for the development of more cutting-edge bioinformatics tools to help with data analysis, and called for collaborations between bench researchers and bioinformatics specialists to develop those tools. Dr. Ransohoff acknowledged that bioinformatics is important, but pointed out that many of the problems in data analysis and interpretation of the omics field are not new problems stemming from the nature of the technology. Instead, they are age-old problems well known to clinical epidemiologists: overfitting of data, bias, and sample sizes that are disproportionately small compared to the number of variables measured. Researchers in the omics field do not pay enough attention to these experimental design flaws that can distort the accuracy and reproducibility of their results, Dr. Ransohoff said.

Overfitting of data is a problem in a number of omics studies, Drs. Quackenbush and Ransohoff asserted. Overfitting can occur when a large number of predictive variables are fit to a small number of subjects. A model can fit perfectly by chance in these situations, even if there is no real relationship, Dr. Ransohoff pointed out. He cited a study by Richard Simon[1] in which Dr. Simon simulated a genomics study by making up patients, assigning them genes with various degrees of expression, and randomly assigning whether or not they had cancer. Dr. Simon then did a multivariable analysis to see if he could find a genetic signature model that discriminated between patients who had cancer versus those who did not. He found that, depending on how he did his analysis, he could make a discrimination model fit the data almost perfectly (98 percent of the time). He was able to achieve high-accuracy assessments of predictive genetic

---

[1]Simon R, et al. 2003. Pitfalls in the use of DNA microarray data for diagnostic and prognostic classification. *Journal of the National Cancer Institute* 95(1):14-18.

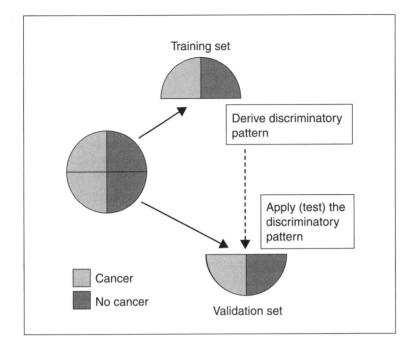

FIGURE 3 Method of dividing original sample to assess reproducibility and overfitting.
SOURCE: Ransohoff presentation (March 20, 2006).

signatures, even though they did not truly determine which patients had cancer, as this was randomly assigned.

The way to check for overfitting is to assess the reproducibility of the results in a new group of subjects who are totally independent from the original group (Figure 3). But such assessments often are not done in omics studies, according to Dr. Ransohoff. Instead, results from a new group of subjects are often combined with those from the original group to further assess the accuracy of a predictive genetic signature or proteomic pattern.

Having a large enough sample population can help avoid the problem of overfitting, Dr. Quackenbush noted. "If we find a biomarker, or a set of patterns that we use as a biomarker, in 20 to 30 samples when we're looking at tens of thousands of genes, there's a high likelihood that when we go to a larger population, many of those genes that we see in the small sample set won't hold up as robust markers," he said.

According to Dr. Ransohoff, overfitting helps to explain why a number of studies of cancer biomarkers, including a Dutch study that recently reported a gene expression signature as a predictor of breast cancer survival,[2] showed initial highly promising results that did not hold up quite as strongly when researchers tried to duplicate them in different study populations. One reanalysis of the original data from seven RNA expression and cancer prognosis studies[3] found that in five of them, results were no better than chance. Dr. Ransohoff pointed out that many of these studies were conducted at well-respected institutions and published in major journals, such as *Lancet* and the *New England Journal of Medicine.* "If our best institutions don't know when the data are strong enough to support claims like this, then there's something genuinely difficult about the entire field," he said.

Dr. Ransohoff said much of the faulty study designs of omics research, and their readily accepted findings by major journals, stems from a culture clash between bench scientists and clinical researchers. "A culture clash hinders exploration when you get people from these fields in the same room and they really can't communicate with one another because the molecular biologists don't understand enough about clinical or observational epidemiology and biostatistics, and the epidemiologists and biostatisticians may be intimidated and don't know enough about molecular biology and biochemistry," he said.

Bias is another common problem in experimental research that is not addressed adequately by many in the omics field, according to Dr. Ransohoff. Bias is the systematic difference between compared groups that alters the accuracy of the conclusions stemming from the comparison. Bias is such a common and serious problem in research that "results of a study must be regarded as being guilty of bias until proven innocent," he said. Just one bias can be a fatal flaw in a study.

As Dr. Quackenbush noted, "I've looked at people's datasets, even published datasets, where they claim differences between two groups, and when I look at it, I see the primary difference being the difference between two hospitals or two collection protocols." As an example of bias, Dr. Ransohoff reported on the reanalysis of the data from studies of the highly acclaimed

---

[2]Van de Vijver MJ, et al. 2002. A gene-expression signature as a predictor of survival in breast cancer. *New England Journal of Medicine* 347(25):1999-2009.

[3]Michiels S, et al. 2005. Prediction of cancer outcome with microarrays: A multiple random validation strategy. *Lancet* 365(9458):488-492.

proteomics test for ovarian cancer, which supposedly could detect ovarian cancer in blood serum with near 100 percent accuracy.[4] When statistician Keith Baggerly scrutinized the methods used to assess the accuracy of the study results, he discovered significant nonbiologic experimental bias between the cancer and control groups. He found that the researchers ran their proteomic analyses of ovarian cancer samples on different days than when they ran the same analyses on noncancer samples. Because of mass spectrometer drift over time, this created a bias because a "signal," from the machine, was introduced into one group but not the other, making the proteomics test result invalid.[5]

In clinical research, the bias of baseline inequality is usually avoided easily and effectively by using randomization, but researchers still go to great lengths simply to report that there are no statistical differences in the baseline conditions of the study populations they are comparing. In contrast, the features needed to assess "baseline inequality" are seldom reported in the same detail in much "omics" research. According to Dr. Ransohoff, "the process to deal with bias is routinely ignored by authors, reviewers and editors in omics research." A number of factors could cause bias in omics research, including differences in how samples are collected, handled, and stored, or in how the assay is run. But such details are rarely reported when this research is published, he said. "When I want to find out what's happened in an article, I've got to go to a Gordon conference and take the researcher out for a walk in the woods and interview [him or her] for an hour. But, of course, that's what our method sections are supposed to do," Dr. Ransohoff said. "Our methods sections are failing the scientific community in much 'omics' research."

Dr. Quackenbush also stressed the need for data and methods to be openly reported in a readily accessible fashion so that other researchers can review them and/or compare the reported data to their own research results. Such reviews and comparisons are key to validating particular biomarkers. But to do such reviews and comparisons, researchers need to know the biological characteristics of the study samples, including relevant clinical information, how the samples were collected and analyzed, and what the

---

[4]Petricoin EF, et al. 2002. Use of proteomic patterns in serum to identify ovarian cancer. *Lancet* 359(9306):572-577.

[5]Check E. 2004. Proteomics and cancer: Running before we can walk? *Nature* 429(6991):496-497.

results were. This information is often missing in published journal articles or data published online, Dr. Quackenbush noted.

To counter that lack of information, he and others at the Microarray Gene Expression Data Society created a guide for authors, editors, and reviewers of microarray gene expression papers. The Minimum Information About a Microarray Experiment (MIAME) guide[6] requires researchers to report effectively on their entire process of collecting, managing, and analyzing data so that the data can be reused and interpreted by others. The MIAME guide was published in 2001, and has been readily adopted by several scientific journals as a requirement for publication. The guide has led to the development of standards in other fields, including metabolomics and proteomics, according to Dr. Quackenbush.

Numerous challenges in the reporting of data still need to be addressed, however, Dr. Quackenbush pointed out. One challenge is to develop a standard format for consistently describing and entering clinical data, such as the estrogen receptor status of a tumor sample, into a database so that the information can be accessed easily by others. "A rose by any other name is a rose, you just can't find it in the database," he said. He suggested "carrots and sticks" from research funders and journals to encourage more standardized reporting of data.

Standard data formats are especially needed so researchers can compare genomic, proteomic, and metabolomic datasets to each other. Such cross-domain comparisons will enable researchers to move more rapidly from the discovery of biomarkers to their applications in the clinic, Dr. Quackenbush said. A centralized repository of omics data would be helpful to make such comparisons, he added, but such a repository does not currently exist. Most researchers are not keen on creating an omics database, because such work is considered "blue-collar science," he said. "It's not very sexy—nobody is going to win a Nobel Prize for creating a database, yet bringing such data together and integrating it is absolutely essential if we want to look beyond these demonstration studies that have been done and really do the large-scale clinical studies we'd like to be able to do."

There also is a need to develop tools that can visualize and interpret omics data in a way that is easy for clinicians to access and understand. Otherwise, omics tests will not be readily adopted in a clinical setting. "You don't want to have to send your data off to a statistician in order to tell a

---

[6] *http://www.mged.org/Workgroups/MIAME/miame_checklist.html.*

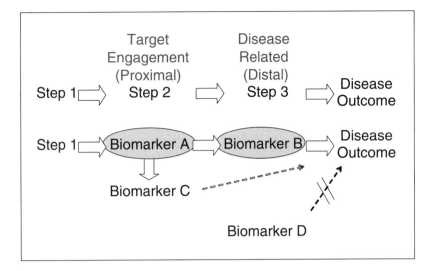

FIGURE 4 Target-engagement markers (Biomarker A) versus disease-related markers (Biomarker B). Pathophysiology is typically a multistep process. A putative biomarker may be (i) involved in one of the steps of the pathophysiology of a disease outcome (Biomarker A), (ii) related to, but not directly involved in, the pathophysiology of a disease outcome (Biomarker C), or (iii) not involved in the pathophysiology of a disease outcome (Biomarker D).
SOURCE: Wagner presentation (March 20, 2006). Adapted from Wagner (2002).

patient whether or not he or she is going to be resistant to chemotherapy," Dr. Quackenbush noted.

Dr. Wagner explored a new angle of biomarker validity in his talk by showing how pharmaceutical companies classify biomarkers and tailor their degree of validity assessments according to the type of biomarker and how it will be used. He began his talk by pointing out how many biomarkers fall at various intervals on the pathophysiology path from the initial trigger or cause of a disease to final disease outcome (Figure 4). Biomarkers that occur close to the actions of the target are termed target-engagement biomarkers. Those that are closer to the disease outcome are called disease-related biomarkers. Target-engagement biomarkers help one understand how well a drug is acting on a target, whereas disease-related biomarkers are used to assess the effect of a particular drug on a disease.

Some biomarkers are not directly related to pathophysiology, yet are still useful. One example is hemoglobin A1c, Dr. Wagner noted. This is a measure of hemoglobin with glucose molecules attached (glycated hemoglobin). When there are higher than normal levels of blood glucose, as occurs with diabetes, more hemoglobin becomes glycated. Blood levels of hemoglobin A1c serve as an excellent surrogate endpoint in diabetes drug trials, yet this biomarker has nothing to do with the diabetes disease process—that is, the glycation of hemoglobin has no impact itself on the health of the patient.

Dr. Quackenbush also pointed out that "there are many examples of biomarkers that exist outside of the realm of omic technologies that are clinically useful even though they don't have a clear mechanistic interpretation." Both prostate-specific antigen (PSA) and carcinogenic embryonic antigen are biomarkers used clinically to manage patients, he noted, but they do not explain tumor behavior. So although finding a mechanistic interpretation can help validate a biomarker, one shouldn't rule out the usefulness of a biomarker if its mechanism of action cannot be directly related to a disease process, Drs. Quackenbush and Wagner pointed out. "If we focus too strongly on just looking at mechanistic understanding in order to develop biomarkers, we may be throwing out the baby with the bathwater," Dr. Quackenbush said.

Another way pharmaceutical companies classify biomarkers is according to the purpose for which they will be used (Box 1). Exploratory biomarkers are usually used to generate hypotheses and are mainly seen as research and development tools. Demonstration biomarkers are considered one step up from that and termed probable or emerging biomarkers, according to FDA parlance.

BOX 1   Biomarker Types

Characterization—known or established biomarker that often aids drug development decision making.

Demonstration—a probable or emerging biomarker.

Disease-related—used to assess the effect of a particular drug on a specific disease process.

Dose stratifier—an indicator of the optimal dose of a specific drug for a specific patient.

BOX 1   Continued

Early compound screening—biomarker used early in drug development to detect likely effective drug candidates, that is, those that affect a specific drug target.

Early response indicator—biomarker that objectively indicates early in treatment whether the patient is responding to the treatment; for example, PET imaging of tumor size.

Exploratory—used to generate hypotheses; a research and development tool.

Partial surrogate endpoint—indicator of the effectiveness of treatment in early (Phase I/II) clinical trials. Improvement of a partial surrogate endpoint is necessary for, but not sufficient to, ensure improvement of the primary clinical endpoint of interest. Partial surrogate endpoints serve as indicators of whether to continue the clinical testing of new drugs and progress to Phase III trials.

Patient classifier—marker that classifies patients by disease subset.

Pharmacodynamic—marker that indicates drug activity and informs dose and schedule selection of a drug.

Relapse risk stratifier—indicator of the degree of risk for relapse after initial therapy.

Response predictor—a measurement made before treatment to predict whether a particular treatment is likely to be beneficial.

Risk management—marker for patients or subgroups with high probability of experiencing adverse effects from their treatment, such as a marker for a drug metabolism subset.

Risk stratifier—indicator of the probability of an event (e.g., metastasis) or time to the event.

Surrogate endpoint—an outcome measure that is thought to correlate with the primary clinical endpoint (outcome) of interest, and is used in place of the primary endpoint to determine whether the treatment is working.

Target-engagement—indicator of how well a drug is acting on a target.

Tumor progression indicator—a measurement that provides early detection of tumor progression following treatment; for example, an increase in PSA levels can indicate progression of prostate cancer.

SOURCE: This box is based on information presented by Drs. Janet Woodcock, John Wagner, and Richard Simon at the workshop.

TABLE 1 Research and Regulatory Use of Qualified Disease-Related Biomarkers

Exploration	Hypothesis generation
Demonstration	Decision making, supporting evidence with primary clinical evidence
Characterization	Decision making, dose finding, secondary/tertiary claims
Surrogacy	Registration

SOURCE: Wagner presentation (March 20, 2006). Adapted from PhRMA Biomarker Working Group, FDA Advisory Committee Meeting (2004).

Characterization biomarkers are known or established biomarkers that often aid drug development decision making, and surrogacy biomarkers can substitute for clinical endpoints in drug efficacy studies.

All biomarkers undergo some degree of validation and qualification. Dr. Wagner defined qualification as the evidentiary process of linking a biomarker with biology and clinical endpoints, generating data that are scientifically and clinically meaningful within the context of its intended use. This contrasts with validation of the biomarker assay, which is obtaining reliable biomarker data that meet the experiment or study objective. The degree of validation and qualification of biomarkers should fit their purpose, and depend upon whether they are target-engagement biomarkers or disease-related biomarkers (Table 1).

Exploratory biomarkers require a minimum set of assay validation experiments, but demonstration or characterization biomarkers require more advanced assay validation. This is especially true if they will be used as a basis for drug development decisions, such as whether a drug is effective, or at what dose the drug should be used. A target-engagement biomarker that is used in drug development decision making would need some advanced validation, but would not be subject to qualification assessments, whereas a disease-related biomarker that would be used for such decision making would undergo qualification assessments, said Dr. Wagner.

COORDINATING THE DEVELOPMENT OF BIOMARKERS AND TARGETED THERAPIES

Only a fraction of cancer patients will respond to a given cancer therapy, with responders being as low as 1 percent for drugs that target specific genetic and molecular changes in cancer cells. Such targeted treatments often require biomarkers that can reliably predict patients likely to respond in order to show efficacy in clinical trials, let alone in the clinical setting at large. But development of biomarker-based tests to predict drug responders has lagged and is often undertaken outside of the company developing the drug. Progress in this field potentially could be accelerated by coordinating the development of biomarkers and new drugs. The goal of the third session of the conference was to discuss current incentives and disincentives for the development of biomarkers for targeted cancer therapies, and ways to encourage cooperation and resource sharing.

Therapeutics Industry Perspective

Paul Waring, PhD, of Genentech opened this session by summarizing the state of the art for developing clinically useful biomarker tests to predict patients likely to respond to targeted cancer therapies. He discussed the first successful attempt in this regard, which was the codevelopment of the breast cancer drug Herceptin with a diagnostic test that predicted whether breast cancer patients would be likely to respond to it. Herceptin targets the gene human epidermal growth factor receptor 2 (HER2), which is overexpressed in about 25 percent of breast cancer cases due to amplification of the gene. Genentech, which developed the drug, also developed an assay to select patients likely to respond for its clinical trial of Herceptin.

Due to the diagnostic test's ability to enrich the study population with drug responders, a clinical trial was able to show that Herceptin lengthened the survival time of about 25 percent of women with metastatic breast cancer who overexpress the HER2 gene. If the study population had not been enriched with responders, a mathematical model revealed the clinical efficacy of the drug would have been difficult, if not impossible, to demonstrate with the number of patients typically recruited for a clinical trial. "This is clearly a huge success that raised the paradigm for personalized medicine and predictive tests in targeted therapies," Dr. Waring said.

The diagnostic assay used to select patients for the clinical trial proved to be unsuitable for commercialization, however, so Genentech partnered with DAKO to codevelop an immunohistochemistry (IHC) diagnostic test, known as the HercepTest®, which is now widely used in clinical practice. This test was validated during the Phase III clinical trial by showing equivalence to the clinical trial assay. Both the drug and the test were approved jointly by the FDA in September 1998.

Studies have shown that there are high false-positive (Table 2) and false-negative (Table 3) rates in the general community for HER2 immunohistochemistry testing as well as the more accurate fluorescent *in situ* hybridization (FISH) test. Although large, more experienced laboratories generally perform both these tests well with low false-positive and -negative rates, small-volume laboratories, particularly those that use home-brew immunohistochemistry tests, were shown in these studies to have unacceptably high false-positive and false-negative rates, Dr. Waring reported. "The problem isn't so much with the tests themselves, but where the tests are performed," he said. Genentech's estimation of the situation is that each year about 5,000 U.S. patients receive Herceptin without any clinical benefit, and about 7,000 patients who could derive benefit are not being treated because of a false-negative test result. "This keeps me awake at night and is a very serious problem," Dr. Waring said.

To rectify this situation, Dr. Waring recommended implementation of standardized testing and mandatory participation in HER2 quality assurance testing programs akin to what is in place in the United Kingdom. Such best practice programs allow laboratories to compare their performance against reference materials and other laboratories and hence identify whether they have a testing problem (Ellis et al., 2004). The accompanying educational material and instructional assistance allows most laboratories to identify and rectify their problems. In the UK HER2 QAP program, which publishes its collective results (Rhodes et al., 2004), retesting of over 100 European laboratories on 6 successive occasions resulted, over a 2 year period, in a significant improvement in the number of laboratories achieving acceptable HER2 test results. Dr. Waring added that "generally, the pathology community isn't ready yet in many ways to adopt these predictive tests for therapeutic decision making. I think for more sophisticated tests, they're going to have to be performed in central reference laboratories that have very rigorous accreditation processes."

Dr. Waring described more recent and less successful attempts to develop diagnostic tests that predict responsiveness to targeted cancer therapies. He discussed the DAKO test for expression of epidermal growth factor receptor (EGFR), which was used to detect colorectal cancer patients likely to respond to cetuximab. Colorectal cancer patients were not entered into the clinical trials of cetuximab unless they had a positive result in the EGFR test (had 1 percent or greater tumor cells showing positivity). These trials revealed that between 10 and 20 percent of patients responded, and led to the approval of the drug by the FDA in 2004.

TABLE 2 False-Positive HER2 Test Results

	Local vs. Central FISH	Local vs. Central HercepTest	Local HercepTest vs. Central FISH	Local Homebrew vs. Central HercepTest
NCCTG N9831 (n=970)	15%	20%	—	31%
B-31 (n=104)	—	14%	21%	—
Small volume (n=79)	—	19%	23%	48%
Large volume (n=24)	—	4%	4%	0%
B-31 amendment (n=204)	2% overall	2% overall	2% overall	2% overall
HER-First (n=1,434)	—	23% (any IHC)	26% (any IHC)	—

SOURCES: Adapted from Waring (2006). Adapted from Perez et al. (2004), Paik et al. (2002), Reddy et al. (2006).

TABLE 3 False-Negative HER2 Test Results

	Local vs. Central FISH	Local vs. Central HercepTest	Local HercepTest vs. Central FISH	Local Homebrew vs. Central HercepTest
NCCTG N9831	15%	20%	—	31%
N9831 (n=970)	—	—	—	—
B-31 (n=104)	—	—	—	—
Small volume (n=79)	—	—	—	—
Large volume (n=24)	—	—	—	—
B-31 amendment (n=204)	—	—	—	—
HER-First (n=1,434)	—	11% (any IHC)	14% (any IHC)	—

SOURCES: Adapted from Waring (2006). Adapted from Reddy et al. (2006).

But an analysis of the trials and other studies has revealed that there is no correlation between clinical benefit and EGFR positivity, either by the number of positive cells or by staining intensity, Dr. Waring pointed out. This is probably because the staining pattern for EGFR is quite heterogeneous, he said. Some tumors may only show focal areas that are positive, so a positive result may depend on which piece of the tumor is examined. "The EGFR test was able to accelerate or increase the probability that cetuximab would be approved and in that regard it was a success. But I don't think it has been a success in terms of testing in community practice," said Dr. Waring.

He also pointed out that although initial studies indicated that more than 70 percent of the responders to Tarceva had mutations in EGFR, testing positive did not correlate with a survival advantage in small cell lung cancer patients because of the complex biology of the disease. Studies have shown that although patients who have EGFR mutations initially respond to these drugs, surviving tumor cells may acquire secondary resistance mutations and then progress, resulting in no survival benefit. But the drug also slows the growth of tumors in patients who do not have EGFR mutations, which can result in improved survival time.

Dr. Waring concluded his talk by suggesting ways to enhance harnessing the power of cancer biomarkers. He recommended designing and powering clinical trials to answer diagnostic questions as well as therapeutic questions. Although large numbers of patients are accrued to clinical trials of cancer drugs, many of their samples are not available or are of inadequate quantity or quality to enable the testing needed to find a molecular signature that correlates with clinical outcome. He also recommended that the clinical utility of predictive diagnostic tests be demonstrated. The test has to significantly impact therapeutic decision making, he said. He also noted the importance of making distinctions between clinical assays used to enroll patients in clinical trials of unproven therapies versus those used to test patients in clinical practice prior to making therapeutic decisions.

Diagnostics Industry Perspective

The next talk was given by Robert Lipshutz, PhD, of Affymetrix. Dr. Lipshutz gave the diagnostics industry perspective on incentives and disincentives to develop biomarker-based cancer diagnostics whose utility is linked to targeted therapies. He noted that companies such as Affymetrix, which develop the platforms used in many microarray diagnostic tests, often partner with drug companies, diagnostic companies, academic institutions, and/or Clinical Laboratory Improvement Amendments (CLIA)-certified laboratories to move a diagnostic test from its initial discovery to its commercial application in a clinical setting. To encourage academic medical centers to undertake biomarker discovery endeavors using their gene testing platform, Affymetrix will offer them special pricing so as to share the risk of pursuing such discovery efforts, as well as the opportunity to develop any biomarkers discovered by working with its partners—companies that will validate and bring the biomarker-based tests to market.

Although the cost of developing a diagnostic is relatively small compared to the cost of developing a drug, so too is the overall return, Dr. Lipshutz noted. "You don't have a lot of blockbuster diagnostics on the market," he said. A number of costs and risks are linked to every step of developing a biomarker-based test, he pointed out. If the diagnostic is only going to be useful if the targeted therapy gains FDA approval, the risk of the new drug failing clinical trials must be added to the risk of developing a new diagnostic. This is a major disincentive for diagnostic companies, he noted.

But on the plus side, if the drug does make it to market with its label requiring the diagnostic test, then there is reduced cost and risk linked to marketing the diagnostic because both are shared with the company that developed the new drug. However, its reimbursement rate may be too low for the diagnostic company to earn enough revenue on the test, even if there is a huge demand for it. Also, Dr. Lipshutz noted that the actual market for diagnostics linked to specific cancer therapeutics is smaller than the average diagnostic test, and thus is likely to generate less revenue. This, too, can act as a disincentive to diagnostic companies.

For simple diagnostics, such as the IHC tests already on the market, the costs and risks were low so it was relatively easy to have diagnostic companies develop these tests, Dr. Lipshutz said. But they may be less inclined to develop more complicated diagnostics that might have to undergo an extensive *in vitro* diagnostics (IVD) approval process with the FDA to reach the market, he added. He explained the IVD process has more extensive testing requirements than the home-brew development process often used for diagnostic tests, which only requires CLIA certification of the laboratory performing the test (Figure 5).

Dr. Lipshutz suggested an alternative regulatory model to reduce the risks and costs of developing biomarker diagnostic tests linked to new targeted cancer treatments. In this model, researchers should use a biomarker

Home-Brew Development Process	IVD Development Process via FDA
• Marker Discovery	• Marker Discovery
• Marker Validation	• Marker Validation
• Assay Development	• Assay Development
• Component Sourcing	• Component Sourcing
• Assay Validation	• Assay Validation
• Documentation	• Kit & Instrument Development
• Test Launch	• Kit & Instrument Validation
	• Shelf Life Studies
	• Multisite Trials
	• Documentation
	• Submission and Approval or Clearance

FIGURE 5 IVD developmental process for FDA approval or clearance vs. home-brew test development.
SOURCE: Adapted from Lipshutz presentation (March 20, 2006).

assay that meets CLIA requirements during clinical trials of the new therapeutic for which its use will be linked. If the drug is then approved by the FDA, the diagnostic test would also enter the market via CLIA-certified labs. But linked to the diagnostic approval would be the stipulation that further testing be done for the diagnostic test so it is later evaluated by the FDA as an IVD.

Dr. Lipshutz concluded his talk by pointing out the need for improved standards for sample preparation and controls for expression reagents, SNPs, and copy number. He also reiterated the need for statistical standards to evaluate the patterns seen in the omics field.

NCI Perspective

The next speaker was James Doroshow, MD, of the National Cancer Institute (NCI), who discussed the agency's goals and funding initiatives in regard to cancer biomarkers. He pointed out a number of new initiatives the agency has undertaken that should further the cancer biomarker field.

One of these is a $100 million investment in a program to develop and test new animal models molecularly engineered to mimic human cancers. These animal models can be used to predict the pharmacodynamics for new cancer drugs, and can ease the development of assays that can predict effectiveness or safety of new drugs in clinical trials. In a later presentation,

Dr. Charles Sawyers, MD, of the University of California, Los Angeles, stressed the usefulness of these animal models in the discovery of genetic signatures that not only indicate promising genetic targets for drugs, but that can be used to test patients for the presence of such targets. He also noted that the predictive power of preclinical models could be transformed by parallel experiments in genetically engineered mice.

The NCI also increased its support of efforts to develop and validate pharmacodynamic *in vitro* assays well in advance of early phase clinical trials. In addition, it recently opened a new laboratory in its Frederick, Maryland facility to develop its molecular toxicology profiling capabilities so as to speed the development of new agents. The agency also has an extensive collection of synthetic and natural products, as well as biologics and tumor and animal cell lines or models that are provided free-of-charge to cancer researchers.

The Institute supports several programs designed to supplement the limited resources in academia to support the transition from molecular targets to drugs. These include the National Cooperative Drug Discovery Group, which is a consortium of about seven or eight academic groups and pharmaceutical companies. Over the past 20 years, this consortium received about $200 million in NCI funds, and in return has generated efforts that led to the approval of five new cancer drugs, including cetuximab.

NCI's Rapid Access to NCI Discovery Resources program develops assays for investigators who submit promising model targets that survive the competitive external review process. The NCI's Rapid Access to Intervention Development program provides the expertise of its staff and additional in-house resources to academic or nonprofit investigators in the extramural community. These individuals compete to have NCI develop their lead compounds into those suitable for submission into clinical trials. Such development may include pharmacology or toxicology studies, efficacy studies in animals, or the formulation of bulk drug. During its nearly 8-year existence, the program has fostered 24 investigational new drug applications at the FDA, Dr. Doroshow reported.

He acknowledged the need for more NCI resources earmarked to supporting the development of biomarker assays, including validation efforts. "It's almost impossible now to get a peer-reviewed grant to develop an assay. That's something we either have to correct in terms of the peer review process, or by doing the assays for our investigators that we work with closely," he said. NCI is currently developing a new program to address this shortcoming, he added.

Clinical Investigator Perspective

Dr. Sawyers gave the clinical investigator perspective on the discovery and development of cancer biomarkers useful in predicting response to targeted therapies. He began his presentation by showing how developments in his lab led to the discovery of a genetic test for predicting resistance to Gleevec or other drugs that target the BCR-ABL translocation in chronic myelogenous leukemia patients. Because this test was simple to develop, it was not difficult to convince a diagnostic company to undertake this project, and the test was launched commercially this past year.

In contrast, when researchers at the University of California, San Francisco, and the University of California, Los Angeles, discovered two biomarkers that predicted response to EGFR inhibitors in glioblastoma patients, the discovery was not readily adopted and developed by a diagnostics company. These biomarkers were more challenging to develop into an assay, according to Dr. Sawyers, because they consisted of two noncommercial antibodies that would probably be quickly outdated by DNA-based diagnostics. Reluctance to develop the assay also stemmed from the likelihood that it would only be applied to the relatively small number of glioblastoma patients, rather than a larger market. Recent movement away from the standard of single-drug treatment for glioblastoma to multiple-drug therapy also made it difficult to confirm the effectiveness of the assay, he added. Because no diagnostics company has developed the assay, only the original discoverers of the biomarkers use them to test glioblastoma patients. Their labs are not really set up to do such extensive testing, Dr. Sawyers noted.

An important deterrent to academic researchers discovering and developing cancer biomarkers is the high cost associated with such efforts, he pointed out. Genomic tests can add more than $1 million to the cost of running a clinical trial, he estimated. "I personally feel it is worth making these investments to do the experiment, but as many of us know, it's not easy to come up with those kinds of funds, even if a trial is actually quite compelling," he said.

Drs. Sawyers and Lipshutz also addressed intellectual property issues. Dr. Sawyers noted that the kinds of information generated from genetic signature analyses are going to be broadly useful because "there will be a limited number of cancer pathways and lots of drugs will be going at these same pathways from different companies and different angles. So there will be a need for a broad base of pathway markers and I see them as sort of precompetitive knowledge." Discovery costs for those pathways and bio-markers should be shared among academia and pharmaceutical, platform, and diagnostics companies, he said. Incentives for commercialization of molecular diagnostic assays must be retained without compromising the need for open access to data, he added. Such open access is critical for meta-analysis of datasets from different trials.

Barbara Weber, MD, a representative from GlaxoSmithKline, noted that her company concurs with Dr. Sawyer's view that biomarker efforts should be precompetitive. Her drug company has released publicly and immediately all its biomarker data in the hopes of encouraging other large pharmaceutical companies to do the same. "The competitive advantage comes from having good molecules that get properly developed, and we can only benefit by making those data publicly available," she said.

In his talk, Dr. Lipshutz discussed how intellectual property uncertainties can act as a disincentive for diagnostic companies to develop tests that may require the licensing of multiple sources of genetic information. For example, one company that uses Affymetrix's microarray platform plans to use a few hundred genes for their diagnostic tests, but they estimate they would have to examine 20,000 pieces of intellectual property patents before pursuing such tests. Dr. Lipshutz deplored the patenting of natural products and natural laws, which he called patenting obvious information. The Supreme Court is currently evaluating one such patent case[7] and its decision will impact the diagnostics arena, he said.

In the meantime, it has been proposed that patent pools be established so there can be "one-stop shopping" to gain access to all the genetic or other such information needed for a diagnostic test. He also suggested the academic community develop more rational economic models and best practice guidelines for the licensing of intellectual property patents.

BIOMARKER DEVELOPMENT AND REGULATORY OVERSIGHT

Biomarker assays are often widely marketed as laboratory services, without FDA clearance or approval. Such assays usually have undergone analytical validation, which indicates the laboratory accuracy of the tests for detecting what they are supposed to detect. But often there are scanty clinical data on predictive value, such as how accurately the tests determine

---

[7]LabCorp versus Metabolite.

a clinical parameter such as disease diagnosis. However, biomarker tests used to screen for or to diagnose cancer, or to develop a treatment plan have considerable potential for harm as well as benefit. As sophisticated biomarker tests that take advantage of the latest developments in molecular biology begin to enter the market, questions have been raised regarding the level of oversight that is warranted for them. The fourth session of the conference explored recent FDA initiatives regarding biomarkers, ways to design new drug clinical trials that use biomarkers, and how biomarkers should be regulated.

FDA Critical Path Initiative

Janet Woodcock, MD, of the FDA opened the session by noting the recent explosion of new scientific knowledge, particularly within molecular biology, and the doubling of funding that biomedical research has received in the past decade. Yet paradoxically, 2004 marked a 20-year low in the introduction of new molecular-entity drugs on the international market, and there has been a decade-long downward trend for new drugs and biologics submitted to be evaluated by the FDA.

To address this mismatch between innovations in biomedical research and lack of a corresponding surge in novel drugs, the FDA issued a white paper in March 2004 called "Innovation or Stagnation: Challenges and Opportunities on the Critical Path to New Medical Products." The paper noted that this mismatch was caused by using 20th-century tools to evaluate 21st-century advances, and that there is a need to apply new science to the tools used to evaluate new medical products. This is especially true regarding biomarkers, the paper pointed out.

Dr. Woodcock noted that despite the hundreds of candidate biomarkers that are published each year, few ever reach a high enough level of clinical correlation to enable decisions in product development or patient management. "Getting that clinical correlation information that we need is very difficult and costly and it just isn't done," she said. "The process for developing biomarkers for various uses is really broken."

She pointed out that a biomarker is not the same as the assay that is developed to analyze the biomarker, and that this assay requires analytical validation. But it is not yet known how to best prove the performance characteristics of a biomarker-based test that employs newer technologies, especially because many lack a gold standard for comparison. She also stressed the wide range of biomarker uses from pharmacodynamic assays to disease diagnosis, and reiterated the need to tailor the qualification package to the biomarker's intended use. For example, an assay used to screen for disease has a much higher bar than a pharmacodynamic assay used in a drug development program. Dr. Woodcock stated that the agency plans to clarify its regulatory acceptance of biomarkers for various uses in future draft guidances.

New trial designs and methods are needed that incorporate biomarkers, especially if there is codevelopment of a diagnostic and a therapeutic, the FDA white paper also pointed out. These trials should use biomarkers that predict patient responders to make the trial more efficient and informative. "The clinical trial process has been highly observational in its conduct, primarily because we don't have the tools to look at the basis for individual response so we look at population responses. But these trials are extremely expensive and it really is important that we get maximum information when we subject human subjects to experiments."

The FDA white paper also called for more development of bioinformatics, which would encourage the sharing of data and databases so that "we can learn generalizable knowledge about biomarkers, rather than knowledge that simply stays in a particular trial or drug development program," Dr. Woodcock said. There should be standardization of terminology to allow pooling of data and construction of computer-based, quantitative disease models in which biomarker performance data can be incorporated for trial modeling and simulation, she added.

Dr. Woodcock ended her talk by describing the public-private consortia the FDA has fostered to support biomarker development. These include the Critical Path Institute. This nonprofit institute is a consortium consisting of pharmaceutical industry partners with the goal of qualifying new animal safety biomarkers for predicting human toxicities. The companies that participate in this consortium share and cross-validate existing proprietary markers and data that are accrued on them. Another consortium that is under way is an outgrowth of the Oncology Biomarker Qualification Initiative created to join the FDA, NCI, and Centers for Medicare & Medicaid Services (CMS) efforts to foster biomarker development. This led to the development of a nonprofit public-private partnership to qualify fluoro-2-deoxy-D-glucose (FDG)-PET scanning as a marker for drug response in non-Hodgkin's lymphoma.

These consortia are vital, Dr. Woodcock asserted, because "the availability of biomarkers is a common good. It is good for patients and clinicians as well as for researchers and medical-product developers. One company, research or funding source is unlikely to have adequate resources to complete the needed work."

Oversight of Diagnostic Tests

Mr. Heller, a partner at Wilmer Cutler Pickering Hale and Dorr, LLP, gave the next talk. Mr. Heller discussed the FDA's role in regulating biomarker tests and explored some recent precedent-setting initiatives the FDA has taken in regard to regulating innovative biomarker-based assays. Mr. Heller began his talk by stating that the FDA has regulatory jurisdiction over all *in vitro* instruments and reagents that are "intended for use in the diagnosis of disease or other conditions, or in the cure, mitigation, treatment, or prevention of disease" in the human population because they are considered devices.

The FDA defines "intended use" as the objective intent of persons legally responsible for labeling a device. In order to determine intended use, the FDA closely considers a device marketer's advertising, labeling claims, product distribution, product websites, and other objective information, said Mr. Heller. The FDA does not regulate *in vitro* devices that are intended for research purposes only. Instead, the sellers of such devices must comply with a labeling requirement that states the product is for research only and not for clinical purposes. But, "the amount of grayness that attaches to research-only status is profound, and it is something the agency has been wrestling with for years," said Mr. Heller. If someone markets a product for research use and is aware that it is used diagnostically, the agency can assert jurisdiction, and regulate the assay as a device. When the agency asserts jurisdiction, this typically results in premarket submissions to the FDA under its premarket approval (PMA)[8] or premarket notification (510[k])[9] requirements.

Mr. Heller discussed how "home-brew" tests, those that are developed by a laboratory in-house for in-house use, present regulatory challenges to the FDA. The FDA, through an exercise of its enforcement discretion,

---

[8]A PMA application usually requires manufacturers to submit clinical data showing that their devices are safe and effective for their intended uses. PMA requirements for diagnostic tests include clinical data demonstrating sensitivity, specificity, and predictive value.

[9]If a product is substantially similar to another legally marketed device that does not require a PMA, it may enter the market through the 510(k) review process. Manufacturers must submit data showing the accuracy and precision of their diagnostic, often including data demonstrating analytical sensitivity and specificity.

has withheld its authority to regulate home-brew diagnostic tests, thus not requiring premarket submissions before their commercial use. Because the data needed for premarket applications are costly and time consuming to procure and assemble, that regulatory treatment appeals to laboratories who devise tests that are essentially in competition with commercially available assays.

Home-brew tests are subject to the regulations of the Clinical Laboratory Improvement Amendments (CLIA), which mandate that each lab create its own performance specification and provide evidence of accuracy, reproducibility, and analytic specificity for the target patient population of a home-brew test. But Mr. Heller emphasized that although the FDA does not regulate laboratories, it asserts that it has jurisdiction to do so, and the CLIA does not displace the Federal Food, Drug, and Cosmetic Act. "The agency's choice not to enter laboratories, I think, represents a resource judgment and a sensitive approach to prioritizing resources," Mr. Heller said.

But this self-imposed limitation of the FDA raises some potential problems, according to Mr. Heller. For example, he noted that laboratories can license intellectual property for home-brew tests to other laboratories. He suggested that "from a public health point of view, there is very little difference in whether the test moves through commerce itself or the IP is licensed and then the test is performed pursuant to a specific recipe, with royalties paid for each test performed." "As things go forward, I think this will present a challenge to [the] FDA and maybe suggests [the need for] a modern means of regulation, including possibly statutory adjustment," he added.

In order to maximize its efficiency in regulating and ensuring the safety and effectiveness of home-brew tests, the FDA regulates commercial analyte-specific reagents (ASRs), which are used to develop home-brew tests.[10,11] ASRs are defined as "antibodies, both polyclonal and monoclonal,

---

[10]"[I]n-house developed tests have not been actively regulated by the [FDA] and the ingredients used in them generally are not produced under FDA assured manufacturing quality control. Other general controls also have not been applied routinely to these products. FDA is not proposing a comprehensive regulatory scheme over the final tests produced by these laboratories and is focusing instead on the 'active ingredients' (ASRs) provided to the laboratories. However, at a future date, the agency may reevaluate whether additional controls over the in-house tests developed by such laboratories may be needed to provide an appropriate level of consumer protection. Such controls may be especially relevant as testing for the presence of genes associated with cancer or dementing diseases becomes more widely available." *Medical Devices; Classification/Reclassification; Restricted Devices; Analyte Specific Reagents, Prop. Rule, 61 Fed. Reg. 10,484 (March 14, 1996).*

[11]Only CLIA certified high-complexity laboratories may purchase ASRs.

specific receptor proteins, ligands, nucleic acid sequences, and similar reagents which, through specific binding or chemical reaction with substances in a specimen, are intended for use in a diagnostic application for identification and quantification of an individual chemical substance or ligand in biological specimens."[12] Laboratories that produce ASRs must register with the FDA and satisfy the agency's Quality System Regulation (good manufacturing practices). However, most ASRs are not subject to premarket review. Mr. Heller noted that sellers of reagents assert that "many products [on] the market are either research-use only or analyte-specific reagents, whether they necessarily meet those clear definitions or not."

Mr. Heller briefly described an instance in which the FDA made a decision to regulate a microarray product as a device based on its intended clinical use despite the manufacturer's characterization of the product as an ASR, which does not require premarket review. Specifically, Roche Molecular Diagnostics planned to introduce a microarray genetic test for drug metabolism (AmpliChip CYP 450) into marketplace in 2003. After reviewing the product and requesting information from the company, the FDA decided that the product was "of substantial importance in preventing impairment of human health" and its technological characteristics "would cause it to differ from existing or reasonably foreseeable ASRs." This determination resulted in denying a 510(k) exempt status accorded to Class I ASRs and resulted in the requirement to submit a premarket notification before marketing.[13] The FDA suggested that if the device were found to be not substantially equivalent, the company could seek *de novo* classification. *De novo* classification became part of the Federal Food, Drug, and Cosmetic Act in 1997 to provide the FDA with a cost-effective means of avoiding an automatic classification of novel devices into a Class III, PMA status. If a novel device has a lower risk profile that permits the device to be regulated in Class I or II, then the agency has 60 days after receiving a request for *de novo* to classify the device.[14] In this case, both Roche's microarray and Affymetrix's scanner used with the microarray were found not substantially equivalent to a predicate device and both were placed into Class II under

---

[12]21 C.F.R. § 864.4020(a).

[13]Letter from OIVD to Roche Molecular Diagnostics Re: AmpliChip, *http://www.fda. gov/cdrh/oivd/amplichip.html.*

[14]In order to be eligible for *de novo* classification, a 510(k) submitter must submit a request to the agency within 30 days of receiving a not substantially equivalent order proposing and justifying a Class I or II classification.

the *de novo* classification procedure. As a result, each was marketed without a PMA.

Mr. Heller noted that the FDA is very interested in molecular diagnostics and is still trying to determine to what extent it will implement its jurisdiction over new diagnostic devices, as indicated by a number of recent FDA activities. He gave several examples where the FDA asserted regulatory authority over products that manufacturers thought would be outside of the FDA's jurisdiction. Mr. Heller described a meeting and letters in 2004 between the FDA and the developer of a new serum protein test that used mass spectrometry for ovarian cancer screening (OvaCheck). After reviewing the product information and corresponding with the developer, the FDA allowed tests to be run in labs under CLIA without premarket review, but it considered the software used to analyze the results to be a device subject to its regulation and requiring premarket approval.

An April 2003 FDA draft guidance (which is non-binding) on multiplex genetic tests states that tests that interrogate several analytes are not ASRs and require premarket submissions. The focus of the document is on nucleic-acid-based analyses, but the guidance also indicates that it is applicable to protein and tissue arrays. Based on this guidance document, the FDA sent a warning letter to the Nanogen Corporation on August 11, 2005, in which it wrote that the Nanochip Molecular Biology Workstation, Nanochip Electronic Microarray, and several ASRs were not approved as a system or as components. The agency was concerned that the NanoChip array, and the system as a whole, would be used in clinical diagnostics, and was therefore not a research-use only product, as the company had alleged. Similarly, that same month, Access Genetics received a warning letter regarding marketing of test packages for several genetic tests. In addition to notifying a company of concerns with its practices, warning letters can be used by the FDA to clarify how it defines boundaries for its regulatory jurisdiction, according to Mr. Heller.

Mr. Heller noted that biomarker tests used to identify likely responders to drugs will be regulated as devices in parallel with their corresponding drug candidates, and those for higher risk conditions will require PMAs. He added that the FDA Guidance on Pharmacogenomic Data Submissions (March 2005) recommends submitting pharmacogenomic data when the data will be used to make approval-related decisions and when the data are relied upon to define, for example, trial inclusion or exclusion criteria, the assessment for prognosis, dosing, or labeling, or used to support the safety and efficacy of a drug. If a test shows promise for enhancing dosing, safety or effectiveness, or will be specifically referenced on the label, the FDA recommends co-development of the device and drug.

In its April 2005 concept paper on co-development, the FDA addressed the use of a single test with a single drug. Co-development applies when use of an *in vitro* diagnostic is mandatory for drug selection for patients, or when optional use during drug development may assist in understanding disease mechanisms and in selecting clinical trial populations. Co-development applies to a device/drug combination product, as well as to *in vitro* devices and drugs sold separately. The concept paper on co-development explicitly states that drug selection biomarkers, particularly for high-risk conditions, are expected to be subject to PMAs.

Mr. Heller concluded his talk by noting that because more than one center at the FDA will often be involved in approval or clearance decisions, the agency should focus on ensuring coordination among its centers to facilitate the clearance or approval of molecular diagnostics. He suggested that the agency should also focus on clarifying which *in vitro* tests are considered research only, the FDA's role in regulating or not regulating labs, the agency's reliance on the CLIA, and what does and does not constitute an ASR. "Frankly, without these understandings, many folks have products out there, some in perfectly good faith, not knowing that, from an agency perspective, they may be in violation of the law," he said. He added that except for the highest risk *in vitro* diagnostic devices, the FDA should seriously consider *de novo* classification as the standard means of clearing novel molecular diagnostics to ensure safety and effectiveness, so that important diagnostics/prognostics reach health care professionals and patients as soon as possible.

Designing Clinical Studies of Biomarkers

The next speaker was Richard Simon, DSc, of the NCI. Dr. Simon focused on ways that biomarkers are transforming the design of clinical trials, and how they should be appropriately regulated. The conventional wisdom is that there should be broad eligibility of patients in clinical trials. But this notion is outdated now that there is increasing evidence that many kinds of cancers are heterogeneous in pathogenesis and sensitivity to treatment. This results in the effectiveness of many drugs being missed in traditional clinical trials because the proportion of patients who would benefit from the drug was too small to make its presence felt among the majority. "I think it is almost the rule, rather than the exception in cancer therapy, that we treat the majority for the benefit of the minority," Dr. Simon observed.

Instead, he noted that enriching trial populations with likely responders not only will reduce the cost of a clinical trial, but will make it more likely that participants will benefit from the drug being tested. "Cancer clinical trials of molecularly targeted agents may benefit a relatively small proportion of patients, but the benefit for the sensitive subset can be very substantial," he pointed out. New cancer drug development, consequently, increasingly relies on a biomarker classifier that selects a target patient population for treatment. However, the focus of a clinical trial that uses a classifier is to evaluate the effectiveness of a new drug, not to validate the classifier, he said.

Dr. Simon gave several examples of how clinical trials could be designed to incorporate a classifier. In one trial design, a classifier that predicts responsiveness is used to restrict the eligibility of patients to a prospectively planned evaluation of a new drug such that only those who "pass" the responsiveness test are entered into the study and randomized into treatment or control groups (Figure 6).

In another trial design, the responsiveness diagnostic is not used to restrict eligibility, but to structure a prospective analysis plan. The purpose of this trial is to evaluate treatment versus control overall, as well as for a predefined subset of likely responders (Figure 7). The purpose of the trial is neither to reevaluate the components of the classifier, nor to modify or refine it, Dr. Simon stressed.

In the second study design, effectiveness of the new drug in patients is compared to results in controls in the overall study population. If statistical significance (p less than .04) is found, one can claim effectiveness of the drug for the eligible population as a whole. Otherwise, one would perform a single subset analysis that evaluates the drug in the classifier-positive patients, and would claim effectiveness for these patients if statistical significance (p less than .01) is found. The overall study type 1 error of .05 is split between the overall test and the subset test.

The second study design is commonly used when there is not complete confidence that the biomarker used as a classifier will predict response, Dr. Simon noted. The key features of this trial design are that it has a prespecified analysis plan with a single predefined subset. "Saying that the study should be stratified is not enough. You really need a completely well-defined analysis plan as to how you are going to use that subset," he said.

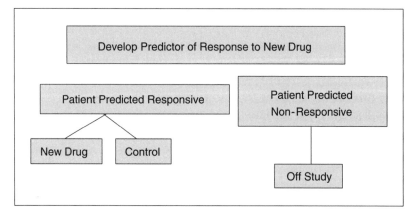

FIGURE 6 Trial strategy I: Utilization of a classifier in developmental strategy for novel drugs.
SOURCE: Simon presentation (March 21, 2006).

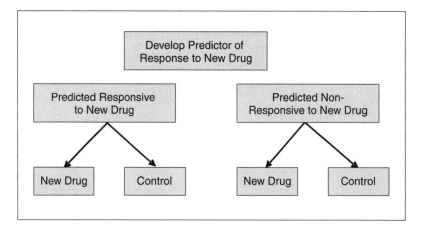

FIGURE 7 Trial strategy II: Treatment response in controls and in predicted responders and nonresponders.
SOURCE: Simon presentation (March 21, 2006).

One can size the trial based on what is needed for the overall analysis. If the results in this analysis are not statistically significant, one could continue accruing for the predetermined subset until a large enough population is reached for a subset analysis. Alternatively, if an interim analysis reveals that there is a large treatment effect for the subset, then one could continue to accrue the classifier-negative patients until there is a large enough population to assess whether the new drug benefits them as well.

A guiding principle for all these study designs is that the data used to develop the classifier must be distinct from the data used to test the hypothesis about treatment effect in subsets determined by the classifier, Dr. Simon pointed out. He added that archived samples from a conventional nontargeted clinical trial could be used to develop the classifier of a subset of likely responders. That subset hypothesis would then be tested in a separate trial. But he noted that it is not possible to use many genetic analysis techniques on archived samples because of the way the samples are preserved.

Dr. Simon concluded his talk by asserting that extensive FDA regulation of biomarkers used in clinical trials is not appropriate. There should be no requirement for demonstrating that the classifier or any of its components are "validated biomarkers of disease status" nor should one have to repeat the classifier development process on independent data, he said. He also does not believe the FDA should regulate how DNA microarrays are used for classifier development in early (Phase I and II) clinical trials.

"If we have developed a classifier in Phase I and Phase II studies, we need to know that we can reproducibly measure that with some assay, and then we need to know something about treatment effect on the subset determined by that classifier. But I don't think it is appropriate to regulate all of the possible ways we could develop that classifier," Dr. Simon said. "[The] FDA can slow effective utilization of this technology, either by overregulating classifier development or by not providing sponsors with a clear and practical roadmap of what is required." He added that some aspects of the FDA guidelines on biomarkers are inappropriate for treatment selection biomarkers.

ASSESSMENT AND ADOPTION OF BIOMARKER-BASED TECHNOLOGIES

Once cancer biomarker tests enter the market, they have to overcome additional hurdles before they are widely used clinically. How readily biomarker tests are adopted in the clinic depends, in part, on how extensively they are reimbursed by health insurers, and how highly they are recommended by various organizations, particularly those that promulgate practice guidelines. Reimbursement policy, in turn, can impact industry marketing and development strategies. The goal of the fifth session of the conference was to examine current and developing strategies for medical decision making and insurance coverage of biomarker-based tests.

Federal Programs for Technology Assessment

Alfred Berg, MD, MPH, of the University of Washington began this session by recounting how the United States Preventive Services Task Force (USPSTF) generates its evidence-based practice recommendations. These recommendations, although not officially binding, generally become the standard of care for medical practice in the United States. A member of the current USPSTF, Dr. Berg explained that it is a rotating, interdisciplinary panel, which regularly publishes its guidelines and recommendations on the web.[15] Its mission is to produce scientific evidence-based reviews of preventive interventions given to asymptomatic patients in primary-care clinical settings.

Prior to conducting their reviews, the Task Force selects a panel of expert generalists who have not already taken public stands on the preventive intervention the panel is reviewing. The analytical framework for the review is specified in advance. It includes assessing how an intervention affects morbidity and mortality, as well as what adverse effects are linked to the intervention, and how the benefits and risks compare to those of standard treatment. The panel does an explicit and prospective quality review of relevant journal articles that meet its stringent criteria. "I emphasize prospective," said Dr. Berg. "We feel strongly that one should specify in advance exactly what you are looking for and not change your mind once you get into the literature."

The reviews are then summarized in evidence tables and the literature is formally linked to recommendations and clinical discussion. Recommendations for interventions that have a net benefit are coded A, B, or C, with the most benefit seen in A recommendations, and the smallest seen in C designations. The C designation is essentially no recommendation because there is fair to good evidence that the benefits and harms are closely

---

[15] *http://www.preventiveservices.ahrq.gov/.*

balanced. Interventions with zero benefit or those that have negative net effects are coded D, and those for whom the evidence is poor are termed I. The quality of the evidence is also considered. To receive an A recommendation, for example, there must be good-quality evidence of a substantial benefit. A substantial benefit seen in a poorly controlled study will not suffice (Table 4).

Many recommendations end up in I territory, Dr. Berg noted. I stands for insufficient because the evidence is insufficient to recommend for or against the intervention. It can be insufficient due to poor quality of the studies done on the intervention, or a lack of studies. An I rating is also given if there are good-quality studies, but their results conflict with each other.

The Task Force recently reviewed the evidence regarding screening for prostate cancer with the PSA test. It gave the use of this test for this purpose a designation of I. Although it found good evidence that screening can detect early stage prostate cancer, there was mixed and inconclusive evidence that such early detection improves health outcomes. In addition, it found very strong evidence that screening and subsequent treatment are both linked to important harms, and concluded that the benefits of treating early prostate cancer are unknown. "The conclusion is not to not do PSA screening," Dr. Berg noted. "The conclusion is that the evidence is insufficient to be able to give clear advice. Our advice to clinicians is that if you are going to do it, do it with care and make sure the patient knows what [he is] getting into."

As Dr. Berg pointed out, the infrequent patient who "wins the lottery" and has a lethal prostate cancer detected at an early stage with PSA screening could receive enormous benefit from such detection. But most screened patients will not receive that benefit. Studies suggest that to prevent one death from prostate cancer in 8 years, one would have to screen about 1,000 men with the PSA. These men would be subject to such potential harms

TABLE 4 Recommendation Codes

Quality of Evidence	Net Benefit			
	Substantial	Moderate	Small	Zero/Negative
Good	A	B	C	D
Fair	B	B	C	D
Poor			I	

SOURCE: Adapted from Berg presentation (March 21, 2006).

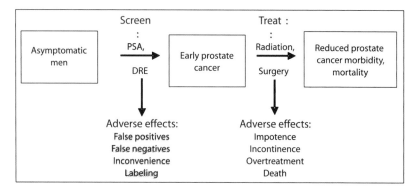

FIGURE 8 Analytic framework for prostate cancer.
SOURCE: Adapted from Berg presentation (March 21, 2006).

as false-positive tests, anxiety, and treatment-linked erectile dysfunction, incontinence, and bowel dysfunction (Figure 8).

"This is a classic dilemma for the patient and the clinician trying to decide whether prostate cancer screening is a good idea for one personally; trying to balance the potential for an enormous benefit against a somewhat more likely potential for harm," Dr. Berg said. The conclusion of the Task Force was basically to let the patient decide whether he wants to receive PSA screening after age 50.

Dr. Berg summed up the findings of the Task Force by noting that their review of biomarker-based tests and other screening tests for skin, bladder, lung, ovarian, pancreatic, oral, and testicular cancer led to I or D recommendations. The only cancer screening tests they actually recommended with B or A ratings were for breast, colorectal, or cervical cancer, and none were biomarker-based tests.

Dr. Berg finished his talk by discussing a new panel sponsored by the Centers for Disease Control and Prevention called the Evaluation of Genomic Applications in Practice and Prevention (EGAPP). Like the Preventive Services Task Force, it is a nonregulatory panel that is expected to make evidence-based recommendations. The goal of EGAPP is to establish and evaluate a systematic and sustainable mechanism for premarket and postmarket assessments of genomic tests in the United States.

"Screening tests are often implemented before the science is fully in place," Dr. Berg noted. "A concern shared by clinicians, patients, regulators, and insurance carriers is that some genomic tests may be released and marketed prematurely. So one of the things that EGAPP hopes to do is to collect what information we do have and assist folks in making a more informed decision."

EGAPP is in the second year of its 3-year existence. It has developed a number of brief summaries of genetic tests. (It chose to review those tests based on health burden of the applicable diseases, and availability, misuse, or impact of the tests.) It is currently working on developing an appropriate analytic framework, as well as a study search strategy and standard for assessing study quality that can be used to review genomic clinical tests. The panel has defined the relevant categories of outcomes for genomic tests. In addition to considering how the test will affect diagnostic determinations, therapeutic choice, and patient outcomes, EGAPP also considers the impact of the test on the families that are related to the person being tested, as well as the impact to society at large.

EGAPP has reviews under way for tests for the drug and toxin metabolizing enzyme, CYP450; the genetic biomarker for colon cancer, HNPCC; and ovarian cancer screening, for which it will be testing its methods. It also plans to do fast-track reviews for tests that have limited data. These include the test for EGFR, and a test for UGT1A1, a drug and toxin metabolizing enzyme that affects susceptibility to chemotherapy side effects. The final expected outcome of the panel is three to five major reviews, two to three fast-track reviews, and a document on methods and evaluation.

Dr. Berg noted that his work on the panel made him aware that there is a lack of information on many important areas related to genetic tests, such as the frequency of genetic variation in the general population, and gene penetrance (what percentage of people with a specific gene allele actually express the allele and show its corresponding phenotype). There are also few clinical trials that compare a genomic intervention with no intervention, and many studies do not assess all the relevant outcomes, he said. Often little attention is paid to documenting the harms of a genomic test, or to its cost and feasibility. Instead, most attention is focused on the potential benefits of a particular test.

Insurance Coverage Decisions and Practice Guidelines

The next speaker was William McGivney, PhD, of the National Comprehensive Cancer Network (NCCN). Dr. McGivney previously was Vice President for Clinical and Coverage Policy of Aetna Health Plans and currently is part of the IOM Medicare Coverage Advisory Committee. He spent much of his talk noting the factors that payors weigh when considering coverage decisions for various diagnostic and treatment interventions, and how those decisions are influenced by societal pressures.

In the early 1990s, pressure from large companies, who wanted to reduce the costs of the health insurance they were providing for their employees, led to the development of strict evidence-based reimbursement decisions, according to Dr. McGivney. But the adverse publicity and lawsuits this approach triggered led to insurance companies seeking other ways of reducing costs, such as reducing how much they reimburse hospitals, physicians, and other health care providers. However, there is still a need to reduce costs and improve health care. Various options are being considered in this regard, including increasing patient copayments and patient participation in treatment decisions, and evaluating and improving the quality of care based on adherence to guidelines and quality measures, Dr. McGivney said.

Biomarker tests present another health care expense that could be a cost challenge for insurers. But their additional cost might be offset by the opportunity to better direct appropriate treatment and derive greater patient benefit for each health care dollar spent. Dr. McGivney noted, "that is the promise of biomarkers, so payors are looking at them as a way to manage and improve utilization and effectiveness by applying them as inclusive criteria even in preauthorization and medical necessity determinations."

When making reimbursement decisions, some payors only consider whether a biomarker provides information that helps manage patients, whereas others also consider what patient outcomes the use of the biomarker improves and carefully examine the evidence in that regard, Dr. McGivney said. Cost effectiveness is not used as a criterion for coverage determinations, he said. But cost does affect the intensity at which a payor reviews the evidence for a reimbursement decision. He noted that "until test kits hit $3,000 and are going to be used in, say, 500,000 patients per year, they may not really care. But at some point, there will be a threshold in terms of dollars, where the payors begin to take a hard look at the impact of the test on their bottom lines."

Dr. McGivney spent part of his talk explaining how many health insurers make their reimbursement decisions. To be reimbursed by an insurer, a technology usually must receive approval from the FDA or some other government regulatory agency. There also has to be sufficient scientific evidence that the technology improves the net health outcome, and must be as beneficial as any established alternatives. The improvement in health benefit this technology provides must also occur outside of a research setting. He noted that the definition of health outcome in cancer is moving away from complete and partial responses to progression-free survival, as the disease becomes more of a chronic condition.

Other unspoken factors also influence reimbursement decisions, Dr. McGivney added. For example, there can be less certainty about the effectiveness of treatments for life-threatening diseases, especially when children are involved. "At Aetna, my unspoken principle was that we paid for everything for kids under 21," he said.

Dr. McGivney concluded his talk by discussing how NCCN guidelines affect clinical care and reimbursement decisions. These guidelines are internationally recognized as the standard for clinical policy and coverage decision in oncology, and are used by CMS and other private payors, he said. They are developed by 1,000 clinicians and patient representatives, who serve on 48 panels focused on individual cancers or supportive care issues. The NCCN guidelines are current, specific, and continually updated, according to Dr. McGivney.

Like the recommendations given by the USPSTF, those given by NCCN specify the level of evidence and consensus. Biomarker tests are increasingly being included in NCCN guidelines, Dr. McGivney noted. "Biomarkers clearly address the direction of each treatment pathway for individual patient subpopulations," he said. Some cancer biomarker tests, such as those for HER2 or the estrogen receptor, play important roles in NCCN guidelines for the treatment of breast cancer (Figure 9). Others, such as urinary urothelial tumor markers, are considered optional additions because the evidence for their effectiveness is not as strong.

Sometimes an NCCN recommendation may contradict what is specified in an FDA label. For example, NCCN recommends that no patient be included or excluded from cetuximab therapy for colorectal cancer on the basis of EGFR test results. In contrast the FDA label for this drug specifies that it be used for the treatment of EGFR-expressing colorectal carcinoma. The decision to link EGFR test results to cetuximab use on the drug label was based on the limited available evidence at the time, and may have also entailed political considerations, Dr. McGivney said.

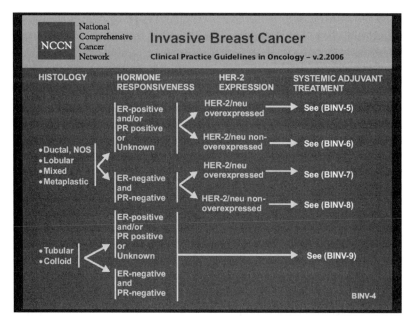

FIGURE 9 National Comprehensive Cancer Network (NCCN) clinical practice guidelines in oncology, v.2.2006.
SOURCE: McGivney presentation (March 21, 2006).

CMS Coverage of Biomarkers

The final talk in this session was given by Jim Rollins, MD, PhD, of CMS. He explained that CMS bases its coverage of a new diagnostic test on its accuracy and whether the test will lead to a better health outcome. To assess the accuracy, sensitivity and specificity measures may not be adequate, he added, and instead the agency may focus on the test's analytic validity, clinical validity, and clinical utility. Often CMS will not accept surrogate markers for survival, such as effect on tumor size, he said. Cost is not a consideration when CMS makes its reimbursement decisions, he added.

Certain factors germane to the older population CMS serves (85 percent are 65 and older) and its limited mandate affect CMS coverage of biomarker-based tests, Dr. Rollins noted. The Medicare statute covers diagnosis and treatment but does not specify a benefit category for screening tests or preventive care, so it is unlikely to reimburse for biomarker-based tests that are used to screen for cancer and/or predict cancer susceptibility. That could potentially change in the future, however, as the Secretary of Health and Human Services' Advisory Committee for Genetics Health in Society has recommended that Congress add a preventive health benefit to areas that are currently being covered by CMS, according to Dr. Rollins. CMS also is not given authority to conduct research. But CMS can give coverage for a medical intervention conditional on the agency's concurrent collection of data on the intervention while reimbursing it. A guidance document on Coverage with Evidence Development is pending on this matter.

Cancer biomarkers used to monitor or manage the care of patients with cancer, including those that predict recurrence, are usually covered by Medicare. For example, the agency on a national level covers the use of CA-125 for peritoneal and ovarian cancer patients. Locally in California, CMS covers the use of the OncoTypeDX test, a genetic test that predicts breast cancer recurrence.

Ninety percent of CMS coverage decisions are made locally, but national decisions take precedence over local ones, Dr. Rollins explained. Vendors, physicians within CMS, or those in private practice can request local or national coverage decisions. If there is a great deal of inconsistency between regions over coverage of a particular intervention or test, the agency may evaluate it and generate a national coverage decision.

The CMS and the Agency for Healthcare Research and Quality recently reviewed the literature on biomarkers for cancer as to how they are used (whether for diagnosis or for the management or monitoring of patients) and how effective they are for certain forms of cancer. This review will be posted on the CMS website, and eventually will be expanded with accuracy assessments, according to Dr. Rollins.

CMS may require more evidence to cover a biomarker test than would be required by the FDA for the test's approval, Dr. Rollins noted. That is because many of the studies submitted to FDA review do not include sufficient numbers of people 65 or older, so their results may not be applicable to the Medicare population. Also, the FDA may approve a particular technology based on the requirement that the vendor will do postmarketing analysis and surveillance. But often vendors do not provide this additional information, so CMS may not cover the use of that technology until there is sufficient evidence to fully evaluate it, Dr. Rollins said.

ECONOMIC IMPACT OF BIOMARKERS

The rapidly increasing cost of medical care is a major concern and has led to a greater interest in the cost effectiveness of medical interventions. The high cost of health care is often attributed, in part, to the adoption of expensive new technologies. These include new targeted therapies for cancer, which, like more traditional therapies, only benefit a fraction of the patients for whom they may be indicated.

However, appropriate patient selection via accurate diagnostic bio-marker tests to predict responsiveness could substantially improve patient outcome and thus increase the cost effectiveness of treatment. Similarly, if biomarker-based screening tests could be developed to detect cancer at an earlier, more easily treated stage, these new biomarker technologies could have a substantial impact on the economic burden of cancer by reducing the cost of treatment, as well as the overall burden and consequence of disease. The goal of the last session of the conference was to examine how the cost effectiveness of biomarker tests and the value of the information they provide affects their acceptance by health care payors, such as insurance companies and CMS.

Cost-Effectiveness Analysis

The first speaker at this session was Andrew Stevens, MD, of the United Kingdom's National Institute for Health and Clinical Excellence (NICE). This organization assesses the value of various medical interventions. Their assessments are used to set the nation's health service priorities. His talk was followed by that of health economist and physician David Meltzer, MD, PhD, of the University of Chicago. Naomi Aronson, PhD, of the BlueCross BlueShield Technology Evaluation Center (TEC), was the final speaker. BlueCross BlueShield provides health insurance for one out of three privately insured Americans. The company uses its TEC's scientific reviews of medical interventions when making reimbursement decisions.

Dr. Stevens began the session by noting the need for having cost effectiveness as the "fourth hurdle" in health care, after safety, efficacy, and quality. Such a hurdle is imperative given limited financial resources and the high costs of innovative treatments. For example, treatment with imatinib (Gleevec) can cost as much as $66,000 per patient, he pointed out. Cost-effectiveness analyses are often used by Great Britain and other nations with socialized medicine to determine how best to ration the health care services it provides, according to Dr. Stevens.

NICE only approves treatments that are both clinically effective and cost effective, although it does give due consideration to notions of equity and innovation. Cost-effectiveness analyses assess the value of a medical treatment by noting its costs relative to its health benefits. That way, one can choose an intervention for which the cost relative to benefit is less than a threshold value. Health benefits are measured with an index called QALY, for quality-adjusted life-years. This index combines measures of quality of life with length of life. In Great Britain, treatments that cost less than $35,000 per QALY are generally approved, whereas those that cost more than $52,000 per QALY are rarely approved. In his talk, Dr. Meltzer noted that the cost-effectiveness threshold for medical interventions in the United States is between $50,000 and $100,000 per QALY.

But cost-effectiveness appraisals have many shortcomings in their methods that can affect their accuracy, all the speakers at this session pointed out. How valid they are depends on the validity of their measures of health outcomes. But that can be adversely affected by basing them on inadequately controlled studies, studies that do not consider the most useful comparators, or studies that are not long enough to truly assess the health outcome of interest. The use of surrogate markers that do not adequately reflect health outcomes can also be a problem. In addition, quality-of-life measures can vary according to subpopulation, Dr Stevens noted, and cost assessments may not be comprehensive enough.

"So there's an awful lot of subjective analysis in these [cost-effectiveness] appraisals, however scientific the documents [used to make them] seem," he said. In her presentation, Dr. Aronson concurred and added that although her center has done some cost-effective analyses for educational purposes, "there isn't any clear science for the cost-effectiveness threshold. I think it is troubling because often we see cost-effectiveness analyses brought to our attention in a lobbying mode by the sponsors of a technology," who claim the technology should be reimbursed because it is cost effective. But Dr. Meltzer pointed out in his talk that despite their limitations, cost-effectiveness analyses were well accepted and broadly used in the biomedical arena.

Dr. Stevens spent much of his discussion elaborating on the experience NICE has had in evaluating or employing various biomarker diagnostics in their assessments of medical interventions. The value of a biomarker depends on what type it is and how it is used, he noted. For example, he considers a test for antibodies to hepatitis C an "exposure biomarker" test for liver cancer.[16] NICE found this biomarker was not useful in determining who should initiate treatment with interferon and ribavirin (as opposed to watchful waiting) because such costly early treatment only increased the QALY from 21 to 22 years. Such exposure biomarkers are not valuable because of their large lead time and low predictive power, he said.

PSA is a useful biomarker for prostate cancer recurrence or prognosis, but NICE called for more clinical trial evidence when evaluating PSA as a screening test for prostate cancer. In addition to the standard measures of a screening test, such as false-positive and false-negative rates, NICE wanted measures of how the test affected patient health outcomes. Even those patients whose biopsies indicate that they are true positives for the PSA test may not develop an aggressive prostate cancer that requires treatment, Dr. Stevens pointed out. This can be problematic because the treatment for prostate cancer has many severe side effects, he added.

NICE accepted the absence of the Philadelphia chromosome in the bone marrow as a useful surrogate biomarker for improved health outcome for patients with chronic myelogenous leukemia who were treated with Gleevec (Figure 10). This chromosome has the translocation that causes the cancer-triggering mutation that Gleevec targets. The agency recommended offering Gleevec to such patients, despite its high cost, because it was much more effective and had significantly fewer side effects than the standard alternative treatment for this type of leukemia (Figure 11).

The final biomarker example Dr. Stevens presented was the use of O6-methylguanine-DNA methyltransferase (MGMT) methylation status[17] in glioma patients to distinguish a treatable subgroup. Treatment with temozolomide in addition to radiotherapy surpasses NICE's cost-effectiveness threshold. But such treatment only in the subgroup likely to respond, as indicated by MGMT methylation status, gives results that suggest it may be cost effective. MGMT methylation status and other response-predicting biomarkers "have the potential to refine disease and therapy and improve cost effectiveness," Dr. Stevens said. But he added that their impact on cost effectiveness depends on whether they induce a cost

---

[16]A hepatitis C infection substantially increases a person's risk of developing liver cancer.

[17]MGMT is a DNA-repair enzyme and its methylation inactivates the enzyme and makes it unable to repair the DNA in tumors damaged by therapy.

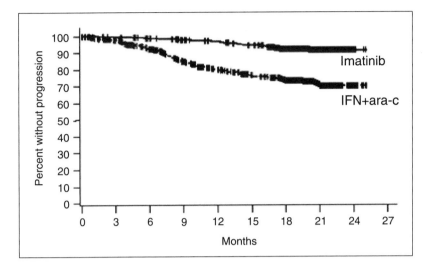

FIGURE 10 Developing new rational therapies—Philadelphia chromosome and imatinib. IFN = interferon alpha; ara-c = cytosine arabinoside.
SOURCE: Stevens presentation (March 21, 2006).

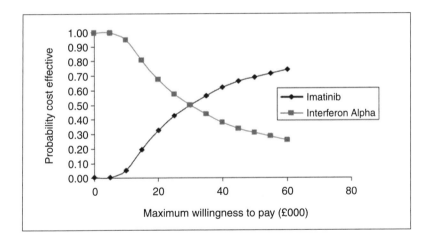

FIGURE 11 Cost-effectiveness acceptability curve for IFN-α and imatinib. The willingness of the National Health Service to pay for a treatment depends on the probability of its cost effectiveness. As cost effectiveness increases, high cost is less of a deterrent to providing the treatment.
SOURCE: Stevens presentation (March 21, 2006).

backlash from drug manufacturers. These companies may increase the price of their drugs to make up for the loss in income due to treatment markets narrowed by biomarker tests for patient responsiveness, he noted.

Dr. Meltzer also pointed out how the value of a diagnostic test, including a biomarker test, depends on how it is used. The cost effectiveness of the Pap test substantially decreases, he showed, when it is used annually or every 2 years, as opposed to every 3 years, because the more frequent use only lengthens a patient's life by an average of a day or two. "These simple analyses can be very revealing," he said. They show that one cannot simply determine whether a test "is good or bad," because such determinations depend, in part, on how the test is used.

He expanded on this concept by showing mathematically how self-selection of a medical treatment by patients occurs because they tend to opt out of a therapy when it is not effective. This self-selection can substantially improve the cost effectiveness of the treatment. But most cost-effective analyses only consider the costs and benefits of a diagnostic or treatment for the entire general population, he noted, and do not consider self-selection. "The results of standard cost effectiveness analyses can be very misleading because in modeling, self-selection is very important," he said.

Requiring copayments for treatments increases self-selection, which in turn also increases the cost effectiveness of the treatment, he added. This suggests a framework for designing copayment strategies to enhance the cost effectiveness of therapies, he said. Nonetheless, he noted that reimbursement systems are not necessarily the right tool to increase value. Decision aids, for example, might be a better tool, he pointed out.

Biomarker diagnostics would be valuable if they encouraged the selective use of treatments. This would substantially increase the cost effectiveness of the treatments, Dr. Meltzer noted. "Our efforts need to go toward getting the right treatment to the right person," he said. "Having a framework to account for heterogeneity in patient benefits is key to valuing diagnostic tests." But he added that "biomarkers can also be used incorrectly in the wrong population. If we use biomarkers outside the context in which they have been developed and use them to find a disease, for example, for which we don't know there is a benefit to treating, then the biomarker is not necessarily going to give us much benefit. So biomarkers are incredibly exciting if they are used right, and dangerous if we don't control the way in which they are used."

Another example of this was the use of COX-2 inhibitors, Dr. Meltzer said. Prior to the release of data showing their cardiovascular side effects,

COX-2 inhibitors were shown to be highly cost-effective drugs for patients at high risk of gastrointestinal bleeding. But the drugs were not cost effective in people at low risk of such bleeding. However, most COX-2 inhibitors were used in the United States by people at low risk of bleeding, so the actual cost effectiveness was poor because of how they were used, Dr. Meltzer pointed out. "We need to think about tests and interventions, not just as they would be used under ideal circumstances, but as they are used in practice," he said.

The Value of Information and Research

In his talk, Dr. Meltzer also showed how mathematical models used to calculate the value of a diagnostic can also be used to calculate the value of information gained by research. These models calculate the difference between the expected outcome with the information garnered from a study and the expected outcome without that information. Such an analysis was used to show the value of research on Alzheimer's disease treatments and wisdom teeth removal that led NICE to invest in such studies, Dr. Meltzer said.

The value of information analysis was also used to calculate the value of biomedical research supported by the National Institutes of Health (NIH). For this calculation, University of Chicago economists showed that biomedical research increases life expectancy in this country by about 3 months per year. By putting a dollar value on that increase in life expectancy for all U.S. citizens, they calculated that biomedical research was worth about $3 trillion a year. This calculation was used to successfully lobby Congress for an increase in the NIH budget, Dr. Meltzer noted. But he pointed out that the real value of research can be far less than expected, in part because it does not always generate the complete information needed to improve a health outcome. For example, he estimated that the expected value of perfect information about prostate cancer generated from research would be $21 billion, but the expected value of more limited information about certain aspects of the disease would be only $1 billion.

Technology Assessment in the Private Sector

Dr. Aronson of BlueCross BlueShield's TEC gave the next presentation. TEC has a staff of physicians, epidemiologists, pharmacists, and medical editors who review and write up scientific assessments of the clinical evidence for various medical interventions. These assessments are used by an independent Medical Advisory Panel, composed mainly of academic researchers, when deciding the insurance company's medical policy, Dr. Aronson explained.

She stressed that the medical policy decisions on which procedures are clinically beneficial are made separately from coverage and reimbursement policy decisions that determine who should receive such clinical benefits and at what rate of reimbursement. In the development of their medical policy, costs and coverage are not considered, Dr. Aronson pointed out, although they are factored into determinations of premium rates, and into the contracts made with health care providers that specify reimbursement rates.

The TEC assessments, some of which are published online at *www.bcbs.com/tec*, consider whether a medical procedure or treatment improves health outcomes by increasing the length and/or quality of life, or by increasing the ability to function. But the organization encounters many challenges when conducting their assessments. According to Dr. Aronson, these challenges include inadequate quality of studies done on a topic, selective reporting and publication bias, and incomplete data from studies that are published; an example is that they do not consider important variables needed to determine medical policy decisions.

Often there is a lack of prospective, randomized, double-blinded, and placebo-controlled clinical studies, Dr. Aronson pointed out. Many clinical studies also lack clearly defined patient populations, relevant comparators, and intention-to-treat analyses of all participants by initial group assignment. The studies often do not have the long-term follow-up needed to adequately assess the health outcomes of a medical intervention. The end result is a lack of robust evidence on the effects of an intervention, and how those effects compare to other interventions, Dr. Aronson said.

The way adverse effects are reported in studies is also problematic, she added. These effects often are not systematically and consistently classified across studies, and are not presented in a way that can be easily synthesized. In addition, usually the frequency of adverse effects, rather than their severity, is reported. "This is tremendously frustrating to us," she said. "We feel like when we do systematic reviews, we can only do half our job. Most of what is available to us is efficacy outcomes, but often, we are really lacking what we need to know about adverse effects."

Another challenge is a lack of direct evidence for the value of diagnostics. Often performance characteristics of the test are used to fill in a model of how the technology can detect a condition or change its management such that there is an improved health outcome. But such an approach can be overly simplistic, Dr. Aronson said, and "there are times when the model is so complicated that you will need direct evidence for a diagnostic; that is to have it tested in a randomized controlled trial, much as if it were an intervention."

For example, there is a test for a variation in the gene that codes for the drug-metabolizing enzyme cytochrome P450 (CYP450). This variant hampers the enzyme's ability to metabolize certain drugs, including warfarin. Therefore, a person who has the gene might benefit by having lower doses of those drugs. But other factors also can slow down drug metabolism. These factors include other enzymes, coexisting disease, age, diet, and interactions with other drugs. Given this complex scenario, it is not clear how useful a test for just one influence on drug metabolism will be for patients who take warfarin, Dr. Aronson said (Figure 12). "I don't think you can jump from the observation about CYP 450 to the conclusion that this is a good control for personalizing warfarin dose. We think something like this is so complicated that it needs to be tested, and I expect many biomarkers for prediction of response may fall into that category," she said.

A final challenge to making assessments of medical interventions or tests that Dr. Aronson discussed was the "file drawer problem" of researchers not publishing their nonsignificant results or results that do not favor the

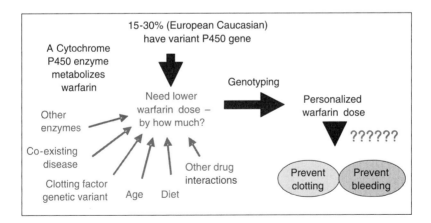

FIGURE 12 Direct evidence for diagnostics.
SOURCE: Adapted from Aronson presentation (March 21, 2006).

drug company that sponsors the research. "We certainly concur [with], and will integrate into our own process, the principles of the International Committee of Medical Journal Editors that call for prospective registration of clinical trials in a public database," she said.

During the next part of her talk, Dr. Aronson addressed some concerns she has about biomarker diagnostics, including their ability to slip through regulatory cracks and not be sufficiently evaluated by adequately designed studies. She noted the regulatory gap for biomarker diagnostics, especially those that are considered home brews. "I see CLIA focusing on lab quality, the FDA focusing on analytic performance, manufacturing quality, clinical validity. Where does clinical utility come in?" she said. She added that she and her colleagues at TEC readily embrace the efforts of EGAPP and believe their model for analyzing genetic tests will serve TEC well.

She criticized the design of many studies to formally assess tumor markers as being inconsistent and inadequate, and questioned the frequent use of biomarkers as surrogates for outcomes in clinical trials. "A correlate does not a surrogate make," she said, because the biomarker may not be in the causal pathway of the disease, there may be multiple causal pathways, or there may be unintended adverse effects of an intervention. It is difficult to know where to draw the line on when a biomarker can adequately serve as a surrogate. She asked, "what shall we trust to demonstrate a health outcome? Under what circumstances? And if we draw the bar there, what are the consequences? Once accepted into clinical practice, it is often difficult to obtain higher level evidence."

Dr. Aronson concluded her talk by discussing cost issues linked to the use of biomarkers. "I think we are, on an ongoing basis, encountering something very troubling in terms of the new technologies—that they bring benefits, but they are small benefits at high costs," she said. Even so-called cost-effective interventions may not be affordable, she added. "I am not sure we can afford everything that is a good buy, or at least at the good buy rate of $50,000 per QALY. Whatever the value of the intervention is, it, alone, cannot ultimately trump affordability," she said.

The consequences of not capping the costs of medical interventions and procedures are high premium rates, which foster a decline in employers offering health benefits to employees, Dr. Aronson said. The end result is that more individuals will be uninsured. The number of uninsured in the United States now is around 45 million—a number that equals the number of Medicare beneficiaries, Dr. Aronson observed.

During the questioning that followed her talk, Dr. Aronson said that although BlueCross BlueShield does fund some health services research, "a research agenda is not really appropriate to our mission nor affordable." The company has a financial responsibility to keep administrative costs, including research costs, to a minimum to maintain premium affordability, she said.

CLINICAL DEVELOPMENT STRATEGIES FOR BIOMARKER UTILIZATION DISCUSSION

On the final day of the conference, representatives from each of the seven small group discussions that met during the previous 2 days gave summaries of their groups' discussions. Discussion moderator Stephen Friend, MD, PhD, of Merck and Co., Inc., started the summary of his group's discussion of clinical development strategies for biomarker utilization by listing the main challenges that his group identified to the clinical development of biomarkers. These challenges were:

- Gaining more access to patient materials and data;
- Coordinating the development of diagnostics and treatments;
- Providing incentives for diagnostic companies;
- Developing "smarter" clinical trial designs; and
- Better integrating basic science and clinical research efforts.

Patient biopsy tissue and other patient materials collected during clinical trials are invaluable for researchers trying to discover or develop biomarkers. But a number of issues contribute to making a lack of patient materials a limiting factor in biomarker clinical development, Dr. Friend said. There is a general lack of sample collection, which is especially true for patients having relapses of their cancer. In addition, the samples are often of poor quality, are misclassified by pathologists such that normal cells are mistaken for cancerous cells, or are preserved in a way that precludes their use in various biomarker studies. Furthermore, annotating and storing samples can be costly, so few investigators or institutions are willing to undertake these endeavors and then provide the samples and data to others.

Even if good-quality patient samples can be accessed, investigators may not be able to use many of them because of inadequate data about the patients from whom they were collected. Variability in the way the data are entered and categorized in a database can also make it difficult, if not impossible, for investigators to retrieve the information they need to include the patients in their studies. Another major barrier that can impede research on stored patient samples is a lack of informed consent forms that are broad enough to encompass new uses of the tissue samples beyond the use for which they were initially collected. In addition, rules in various academic institutions may restrict the sharing of patient materials and corresponding data with other researchers because of competitive and financial pressures. These patient materials are often seen as having some inherent intellectual property value.

Group members had a number of ideas for how to overcome patient sample-related barriers to biomarker development. Suggestions included reexamining the existing informed consent process and making it more flexible as to the range of studies that can be conducted on the tissues collected, providing more funds for the annotation of collected patient materials, and reexamining current academic center rules on collected data, particularly as related to intellectual property. Some members of the discussion group suggested imposing penalties on investigators or academic institutions unwilling to share patient samples and data, whereas other members thought that offering rewards for such behavior was a better alternative.

Some of this group's discussion focused on how to provide incentives for diagnostic companies to translate a biomarker discovery into a clinically marketed test. Dr. Friend noted that diagnostic companies have small profit margins that limit their willingness to undertake major financial risks or costly development endeavors. But a diagnostic company might be more willing to pursue a biomarker test if it could be coupled with a therapeutic that a drug company is developing such that the risks, revenues, and products are shared between the two companies.

Another way to make diagnostic biomarker development more appealing is to give efforts in this regard "type two patents." Australian political philosopher Thomas Pogge coined this term for patents that reward research and development work that results in useful drugs or diagnostics that normally would have a low margin of return. The financial reward of the patent is based on how much of a health benefit the drug or test provides, and makes up for the more limited revenue gained from the sales of the product. Federal governments create funds to support such a patent reward system, which helps make the early development of diagnostic biomarkers less risky. In this manner, the financial worth of a diagnostic is based more on its value to society rather than on its sales revenue.

One member of the group, Richard Frank, MD, PhD, of GE Healthcare, suggested another way to provide an incentive for diagnostics companies to do more research and development on biomarkers. He proposed that these companies be allowed access to government-funded tissue and databanks to conduct biomarker research with the option of having exclusive rights to any biomarker tests that evolve from such research. But other members of the group questioned the need to grant such exclusive rights. Current NIH policy does not grant exclusivity in the licensing of tests that emerge from research on its tissue samples or data.

Another way to make diagnostic development less risky, the group suggested, would be to develop biomarkers for key steps along the biochemical pathways that cause various types of cancer. Researchers suspect there are a limited number of these pathways, which play a role in a wide range of cancers. "If you got those pathway biomarkers, then they are not dependent on a particular individual drug that is going through the pipeline, but instead could apply to any company drug," Dr. Friend said.

He noted that there needs to be a certain level of rigor to a response-predicting diagnostic used in a clinical trial. But the development of such a rigorous biomarker diagnostic often lags behind that of a related drug, so the diagnostic is not ready to enter Phase III testing at the same time as the drug. To solve that timing issue, group participants suggested the development of common shared databases that can be used to develop biomarkers appropriate for predicting response to drugs. They also suggested that industry do more precompetitive investing in research on biomarkers that predict drug response, and make greater use of pathway biomarkers. The FDA might also consider linking its approvals of therapeutics to related response-predicting diagnostics such that one is contingent on the other.

Members of the group also recognized the dynamic nature of the field of biomarkers that predict drug response. Some of these biomarkers are developed after the drugs they predict response to are already on the market, while others are found to be predictive for more drugs than the ones whose labels specify their use. This can restrict the use of biomarker tests because off-label use is often not reimbursed. Consequently, the group suggested considering the need to have more dynamic ways of modifying drug labels based on emerging data.

Group members also suggested considering the consequences of increasingly tight restrictions on off-label use of diagnostics and therapeutics. Not only do such restrictions limit the use of already developed biomarkers, but they limit the amount of resources that drug companies can devote to developing biomarkers by requiring them to sink large sums of money into conducting clinical trials for added indications, Dr. Friend said. However, some group members noted that a benefit to label restrictions could be that more patients would enter clinical trials in order to be reimbursed for a drug used for an off-label purpose.

Another major topic of this group's discussion was the need for smarter clinical trial designs that invest in earlier use of biomarkers, especially to define responsive subpopulations prior to Phase III trials. One person in the group noted that the bulk of a company's drug development costs are for developing unsuccessful drugs—those that do not "pass" clinical trials and enter the market. The use of biomarkers to enrich the number of responders in a clinical trial population should therefore lower development costs overall, he noted, if it makes it more likely that drugs would fare well in clinical trials.

Group participants suggested there should be better integration of basic science and clinical research efforts. The personalized medicine approach that the latest findings in molecular biology suggest does not fit into the traditional models for running clinical trials and developing therapeutics or diagnostics. A "third culture" is needed to bridge the gap between the basic and clinical world, as well as to connect academic and industrial realms, according to group member Dr. Phelps.

Dr. Friend noted that biomarker validation is a "no-person's land" with respect to funding and effort. Academics are not likely to take on this endeavor because they cannot build their careers on such efforts. Pharmaceutical companies also may not be willing to undertake certain biomarker validations if the tests limit their current market for drugs.

Members of the group suggested that NCI support a program that brings together basic, clinical, and perhaps even industry researchers working on a common group of biomarkers—those that define particular oncogenic pathways, for example. This program could be modeled after the Specialized Programs of Research Excellence (SPORE), which supports a mix of basic and clinical researchers working on the same cancer type.

However, David Carbone, MD, of Vanderbilt University, cautioned against emphasizing pathway-specific as opposed to disease-specific research because "the requirements for biomarkers—sensitivity, specificity, precision and accuracy—are quite different in different diseases." He gave the example of epidermal growth factor receptor- (EGFR-) targeted drugs for lung and breast cancer and noted that fundamentally different biomarkers are going to be needed for these two diseases. Dr. Friend agreed with the importance of pathways being seen within the context of what tissues they operate in, but added "it does not take away from the need of taking raw data, aggregating it, and looking at pathways."

Dr. Friend concluded his summary with the group's idea that demonstration studies be funded for oncology drugs already on the market. These studies could demonstrate the feasibility and utility of developing robust response biomarkers.

STRATEGIES FOR IMPLEMENTING STANDARDIZED BIOREPOSITORIES DISCUSSION

This discussion group session was summarized briefly by Harold Moses, MD, of Vanderbilt University. His summary was supplemented with a more detailed written synopsis by Maria Hewitt, PhD, of the IOM.

Dr. Moses noted that the discussion session began with a report by Carolyn Compton, MD, Director of NCI's Office of Biorepositories and Biospecimen Research. She pointed out that NCI's initiatives to further personalized medicine all depend on human biospecimens. Through an extensive internal and external review process, NCI has identified major biorepository-related barriers to furthering these initiatives, as well as potential solutions. This effort led to the development of NCI Guidelines for Biorepositories,[18] the second generation of which is currently being developed in collaboration with the College of American Pathologists and other relevant extramural groups.

The first-generation guidelines include recommendations for the following:

- Common best practices for research biorepositories
- Quality assurance and quality control programs
- Informatics systems
- Ways to address ethical, legal, and policy issues (e.g., informed consent, privacy, data security protections, Institutional Review Board oversight, ownership of and access to biospecimens and data)
- Standardized reporting mechanisms
- Administration and management structure

---

[18] *http://biospecimens.cancer.gov/index.asp.*

The NCI's second-generation guidelines will propose evidence-based standard operating procedures (SOPs). There is widespread recognition that one size does not fit all, and that SOPs may vary depending on the analytic goal.

NCI recently established an Office of Biorepositories and Biospecimen Research and launched a pilot test of the proposed National Biospecimen Network (NBN). This pilot study will be conducted in 11 prostate SPOREs to evaluate the use of best practices. Dr. Compton reported that NBN's key requirements for a new biorepository system include:

- Representation of all cancer types, and all populations
- Access through a timely, centralized peer-review process
- Ethical and privacy compliance through a chain of trust
- Resources provided without intellectual property restrictions
- Pathology and clinical annotation (including longitudinal)
- State-of-the-art information technology system to streamline the research process
- Communication and outreach efforts
- Best practice- and data driven-based SOPs to enable reproducible and comparable (additive) results

Brent Zanke, MD, PhD, of the Ontario Cancer Research Network then discussed the Ontario Tumour Bank, which collects, stores, and distributes tissues at six clinical centers that follow defined SOPs. The Canadian biorepository has centralized data collection, which protects patient privacy, and accrues 3,000 samples a year. A web-based system[19] allows researchers to browse the central database for specimens that meet their study requirements. The Ontario Tumour Bank offers the following products:

- Fresh frozen tumor
- Frozen plasma
- Frozen buffy coat
- Paraffin-embedded tumor
- Normal tissue adjacent to tumor samples

Future plans are to offer paraffin sections, stained sections, and tissue microarrays. Researchers can retrieve extensive data, including specimen

---

[19] *http://www.ontariotumourbank.ca.*

quality, from the web-accessible database on available specimens. Requests for specimen access are made through a controlled application process. A tissue ethics committee oversees the program. Samples are provided at a discount to participating centers and academic researchers in Ontario, and at reimbursement costs to others.

The Ontario biorepository was established with a $10 million (Canadian) investment from the Ontario Provincial Government Ministry of Research and Innovation. Similar biorepositories operate in Great Britain and other countries in Europe, according to Dr. Moses. The Ontario system is not directly exportable to the United States because the United States lacks a national health care system and centralized control of hospitals and provider networks. "Our country is way behind," Dr. Moses said. "The reason the Ontario Tumour Bank can do it for $10 million is that their surgeons, pathologists, etc., are on government salary."

These presentations led to a general discussion on how to fund biorepositories in the United States. Group participants noted that NCI alone cannot bear the costs of supporting national biorepositories, and suggested public-private consortia as a means for supporting biorepositories. Industry has much to gain and should find that sharing costs serves its interest along with academic centers and philanthropists, participants pointed out. The biorepositories would require a large initial investment, Dr. Moses noted, but could be self-sustaining through fees charged for providing high-quality material. Group members also suggested involving CMS and other health care payors in a discussion of supporting biorepository efforts, as they could reimburse pathologists' fees for processing specimens.

In a discussion following Dr. Moses's presentation, David Parkinson, MD, of Amgen (now at Biogen Idec) noted that the biorepository set up by the Multiple Myeloma Research Consortium could serve as a model. This organization is funded by philanthropy and aims to accelerate the development of novel, cutting-edge treatments for multiple myeloma by catalyzing, promoting, and facilitating collaborative research between industry and academia. The Consortium shares its well-annotated and extensive tissue collection with academic and industry researchers. These investigators normally would not focus their efforts on such a rare type of cancer, but do so because of the ease with which these materials are made available to them, Dr. Parkinson said.

Discussant Margaret Spitz, MD, MPH, of the MD Anderson Cancer Center at the University of Texas explored the unique needs of population-based registries and epidemiologic studies. Specimens from control subjects are very important, as are prediagnostic specimens. In addition to clinical information about samples, epidemiologists need information on environmental exposure, family history, and risk factors. Deidentification of specimens can be problematic because long-term follow-up is often necessary. Some large cohort studies have lost funding for longer term follow-up and face the problem of what to do with their patient samples. A plan is needed prospectively to deal with this issue. NIH might support biorepositories as part of large cohort studies, the group suggested. "We need to look at mechanisms for funding the preservation of these biorepositories as a matter of course because they are just too valuable to let deteriorate because of current funding issues," Dr. Moses said.

In a discussion that followed Dr. Moses's presentation, Dr. Carbone described the long-term expenses involved in supporting a biorepository. "It is not readily appreciated how complex managing a tissue collection program really is if you want to do good science," he said. "And it does not end with plunking the sample in liquid nitrogen. The most valuable thing you have in these tissue banks is detailed clinical information that evolves over time. We have to go back every 3 months and go over every sample in our tumor bank and update the status on patients, including what chemotherapies they got. It is very difficult and expensive for very focused questions in a typical SPORE grant with a very limited tissue collection. In one disease site we are talking about it costing between $250,000 and $500,000 a year. The cost of doing anything on a grand scale would be enormous." Dr. Moses noted that NCI currently spends about $50 million a year supporting biorepositories.

A major topic of his discussion group was patient-informed consent. There was an appeal for improved and standardized consent forms that could be used nationally. These forms should resolve current disparities in government rules regarding informed consent, the group suggested. For example, certain government agencies require informed consent before using tissues from patients who have died, while others do not have such a requirement.

The privacy provisions of the Health Insurance Portability and Accountability Act (HIPAA) also raise some obstacles for biorepositories, especially in the area of needing to acquire new patient consent to gain access to tissues for research other than the study for which the samples were originally collected. How hospitals interpret HIPAA rules also varies widely.

The National Cancer Policy Forum was scheduled to examine the effect of HIPAA on biomedical research at its June meeting,[20] Dr. Moses said.

Who owns patient specimens? This was another major issue tackled by the discussion group, which reported that an answer to this question is currently being decided in the courts. One case is testing whether the investigator or hospital owns patient samples. Some research consortia have clearly specified, in advance, issues related to access and ownership of samples.

The discussion group also touched on the need for common data elements being reported for specimens in biorepositories. In addition, some group members suggested biorepositories invest in electronic medical record systems to facilitate gathering of clinical and other data.

In a discussion following Dr. Moses's presentation, Drs. Carbone, Ransohoff, and Friend questioned the logic of investing in large centralized biorepositories because most studies require specific specimens from specific cohorts of patients, and those specimens must be handled in certain ways. "You cannot make some sample adequate for a particular [research] question just by annotation. We have to be careful not to overinvest in large data repositories until we give some thought to exactly what questions we would be able to answer if we really collected all of the data," said Dr. Ransohoff.

Dr. Carbone added, "A much more valuable way to spend resources would be to dramatically increase funding for biospecimens associated with particular clinical investigations or interventions such as cooperative group trials. If you could dump money in support of biospecimen collections into Phase III randomized trials, instead of getting 20 samples out of 1,000 you could get 600, which would give you a specimen collection that is much more valuable than catching things that are thrown in the trash in surgical pathology."

Drs. Friend and Ransohoff suggested starting a new initiative to support centralized biorepositories by funding a collection of just one or two tumor types and focusing on specific hypotheses to "prove that it works and can be a shining example for others," Dr. Friend said. Dr. Moses agreed and reiterated that NBN is funding such a pilot project in its prostate SPOREs.

---

[20]The meeting proceedings will be published as an edited transcript.

STRATEGIES FOR DETERMINING ANALYTIC VALIDITY AND CLINICAL UTILITY OF BIOMARKERS DISCUSSION

Moderator Dr. Howard Schulman reported on his group's discussion about strategies for determining analytic validity and clinical utility of biomarkers. He noted that the group's comments reflected a widespread belief that discussions on biomarkers should differentiate between qualification (clinical validity and utility) and validation (assay validity). Similarly, distinctions should be made for biomarkers used only by pharmaceutical companies during the initial stages of drug development versus those used in clinical trials that affect clinical decisions. For example, biomarkers used only by pharmaceutical companies to determine if a drug they are developing is acting on its target would not have to undergo scrutiny by the FDA, but more oversight is needed for a biomarker used to stratify patients into responders and nonresponders.

Group members stated that the type of technology used for an assay will influence acceptance criteria for its analytical validation. They suggested that one should consider the context and risk/benefit equation when determining validation and qualification acceptance criteria. In other words, the test consequences determine the standard, and assays that affect clinical decisions should meet the highest standards. The discovery phase should be guided by good science without being encumbered by regulations, Dr. Schulman said. However, quality control samples, platform standards, or proficiency testing of laboratories may be needed when a biomarker test is used to predict patient response or determine dosing.

Discussion participants from all of the various interest groups expressed the view that increased access to and standardization of databases is necessary to advance biomarker science. Often investigators cannot access the clinical information they need to conduct biomarker studies. Even if it is collected, it may not be entered into a database in a way that makes it easy to retrieve and use.

The current FDA regulatory approach to biomarkers, which involves having heightened regulation of biomarkers used for clinical decisions and less regulation for those used early on in development, was generally agreed on by the group, Dr. Schulman said. But there was a lack of agreement on acceptance criteria for diagnostics paired with new therapeutics when the diagnostic is a clinical laboratory test that is subject only to CLIA oversight. Members of the group thought it was not necessary to have rigorous regulatory requirements for response-predicting biomarkers used in the initial stages of drug development. It was pointed out that a biomarker used to stratify patients in a clinical trial is not equivalent to a diagnostic test used for the same purpose on the market, and therefore could be subject to a lower standard. Many group members believed that there is a need for a realistic algorithm for combining the development of therapeutics and diagnostics.

In addition, group members thought incentives for developing diagnostics are lacking because the diagnostics business is not a high-margin business. There was tremendous enthusiasm from different interest groups for a variety of consortia that can further precompetitive work on biomarkers. In the discussion after his presentation, Dr. Schulman noted that a consortium for biomarker validation that includes the FDA, CMS, and various pharmaceutical companies is already under way. He added that "there is a general feeling that there is an opportunity to do something right on a bigger scale where oftentimes the intellectual property issues are not problematic."

Group participants noted that investigators who discover biomarkers often do not understand what is required for analytical or clinical validation. "Most people in discovery sites are not familiar with the whole process that one has to go through if you are a diagnostic company," said Dr. Schulman. "It is quite rigorous and for some of these diagnostics you actually have to have data on 6,000 patients whereas oftentimes, people do a study on 20 patients and think they have discovered a diagnostic and are ready to put it out there."

There is a critical need for standards for the new technologies used in biomarker-based assays, Dr. Schulman reported. Standards would be helpful for microarrays and other genomics technologies, proteomics, and metabolomics. These standards could help solve current problems with interpreting results, such as how to deal with uncertainties in protein identification when using mass spectrometer data and how to compare results garnered from different technologies. Another issue is how to establish consistent standards for the same technology, such as for mass spectrometry, whose resolution and other determinants depend on the exact instrumentation used in a laboratory. Even established genetic probes can generate uncertain results when applied to microarrays, Dr. Schulman pointed out.

Members of his discussion group thought there was a role for the National Institute of Standards and Technology and the FDA in resolving some of these standardization issues. Consortia could also be helpful in this regard, he added. In the discussion that followed his presentation, Helen

Francis Lang, PhD, from Affymetrix, Inc. pointed out that "the FDA and a large number of stakeholders are involved with platform companies in establishing quite strict standards in terms of the use of microarrays, controls, and interpretation of data."

In regard to the discovery and development of biomarkers, group members noted that biomarkers are best vetted and promoted to greater degrees of qualification as more studies are conducted with them. "In a way, it is a communal process," Dr. Schulman said. "This is one of the reasons for sharing both samples and information because the more people work on the same set of biomarkers, the more we learn about their flaws and good points and come up with a better test." To further that communal process, the group suggested that investigators publish all raw data from their biomarker studies.

Group participants also noted that better access to clinical specimens would be a boost to diagnostic development, but a number of obstacles must be overcome. As other groups have pointed out, HIPAA can make it difficult to access clinical material and linked clinical information. To overcome those difficulties, members of Dr. Schulman's discussion group suggested facilitating studies by multiple groups that use the same high-quality tissue repository, akin to what is already done successfully in Germany and Holland. These repositories should have extensive annotation of their samples so that the clinical characteristics of interest to investigators are documented. For these samples, much more is needed beyond healthy and diseased distinctions, Dr. Schulman noted.

For biorepositories to be useful, they should be tailored to the types of research investigations that will be done when using them, he added, and contain the kinds of samples and patient information needed to test specific research hypotheses. "For the millions of samples that are stored today in various banks in the United States, most are never touched," Dr. Schulman noted. "That is one reason to link the hypothesis that needs to be tested to the samples collected. If you just collect a priori you may not have the right samples to address your questions." The quality of tissue collections is also critical, so members of the discussion group suggested involving clinical pathologists in the collection process from the start, as well as training them so that sample preparations better meet the needs of researchers.

STRATEGIES TO DEVELOP BIOMARKERS FOR
EARLY DETECTION DISCUSSION

Scott Ramsey, MD, PhD, of Fred Hutchison Cancer Research Center gave the synopsis for his group's discussion on strategies to develop biomarkers for early detection. He noted that most members of his group were "end users" of biomarkers, including clinicians such as oncologists and general internists. They focused their discussion on the use of biomarkers to screen for cancer.

Participants in this group noted that often biomarkers are developed without extensive thought about how exactly they will fit into the clinical pathway, such as how they will affect clinical decisions on treatment or other interventions. "When developing biomarkers for early detection, we must keep end use and value in mind," Dr. Ramsey said. "I have been asked to do clinical studies on biomarkers and when I ask the developer to draw out a decision tree or a pathway and tell me where the biomarker fits in that pathway, I often get four or five different approaches. It is pretty clear that they are looking for whatever sticks in terms of how the biomarker might be used in the clinical pathway."

Biomarker tests are not useful, for example, if they detect a cancer somewhere in the body, but lack enough specificity to lead to a treatment plan. "There is nothing I can imagine worse than having a blood test that shows cancer and not having the foggiest idea what to do with that information as far as the patient is concerned," said discussant Larry Norton, MD, of Memorial Sloan-Kettering Cancer Center.

Biomarkers may also detect heightened risk for a specific cancer, when there are no known prophylactic measures to reduce that risk. Biomarker tests that detect whether people are at higher risk of developing a specific cancer, such as those for breast cancer-related mutations in the BRCA genes, are problematic in that regard. As there are no known measures to substantially reduce the risk of breast cancer in women who test positive for these mutations short of prophylactic mastectomies, the clinical usefulness of the test is questionable, some group members asserted.

Another related area that many individuals in the group thought was often neglected by biomarker test developers was the potential for biomarker tests to harm patients. False results, or positive results for a disease that would never have manifested clinically, can cause patients to be overtreated. Group members thought this concern should be addressed early in the development process before biomarker tests are disseminated, marketed, and adopted. Studies should clearly define the risk/benefit ratio of a biomarker test prior to its use in the clinic, Dr. Ramsey said.

Group participants noted that patient perspectives and preferences for screening tests are very influential. Patients may have expectations about a specific test that are overly optimistic, both in terms of what the test can reveal and in terms of whether it is likely to be negative in their case. Many patients do not have a good sense of the risks and benefits of moving down a pathway of testing using biomarkers. This is especially true for tests that predict heightened risk of developing a cancer. Patients may pressure their physicians to give them these tests, even when there is inadequate evidence regarding the risks and benefits of testing. Alternatively, patients may acquiesce to being tested, even though they have no intention of submitting to treatment if a test is positive. Group members noted the importance of recognizing patient preferences when developing biomarker tests.

Biomarker developers should also consider whether there is a clinical need for a biomarker-based test. Does it help doctors and their patients, or does it just complicate their care? For example, a biomarker-based test for colorectal cancer screening may be superfluous because several good tests that accurately detect this type of cancer are already in use, some discussion participants noted. Often development of a biomarker test is driven by advances in basic science and what new techniques are available, but not necessarily by the clinical need for the test, the group noted. Furthermore, a new understanding of cancer etiology due to basic science findings does not always translate into tests and interventions that are helpful to patients.

Group members recognized that the regulatory oversight for biomarker tests is evolving as a moving target. Some participants expressed concern about off-label use of biomarkers for indications other than those for which they were originally intended. This off-label use can be problematic, as it has been for the PSA test. This test was originally developed for prognostic and surveillance purposes for men diagnosed with prostate cancer. But it was rapidly put into practice as a screening test for prostate cancer, despite a lack of evidence on its risks and benefits in that regard. Tests used for early detection of cancer should be assessed for that purpose, the group stated, with some measure of benefit and risk.

In addition to the FDA's regulatory role, group members noted that coverage decisions play a role in determing the use of biomarker tests in clinical settings. Coverage decisions may keep biomarkers that have not been assessed adequately off the market, or may prevent them from being used inappropriately, such as for screening, discussants noted. But they also appreciated the fact that insurers are extremely responsive to pressures from advocacy groups and politicians, and will reimburse for a test when pressured to do so. Much of the screening technology that Medicare reimburses was mandated by legislation, noted Dr. Ramsey.

Dr. Ramsey finished his presentation by stating his group's awareness that translating a promising discovery into a validated biomarker for early detection of cancer is enormously challenging. As members of other discussion groups noted, this translational process requires access to high-quality, highly annotated patient samples collected in a nonbiased way. They suggested biomarker developers use existing, prospectively collected samples, such as those collected as part of the Women's Health Initiative study.

Smaller biorepositories that stem from private collections can also be useful in this regard, Dr. Norton pointed out, but locating those samples can be difficult. To help researchers find the clinical samples they need, he created a "virtual" repository, which catalogs what is available and uses a computer program to match investigators to appropriate specimen collections. This virtual repository was used to delineate the various types of invasive breast cancers based on tumor cells' estrogen receptor, HER2, progesterone receptor, or EGFR status.

Translating biomarkers into clinically useful tests also requires prospective, randomized clinical trials to assess their risks and benefits. These trials are extremely large, lengthy, and costly. In some cases they may not be feasible because of difficulties in accruing enough patients, especially if current practice patterns make it unlikely that people will knowingly accept randomization into a control group. This is especially true if a biomarker test is put into clinical practice before its clinical usefulness is fully assessed. Such premature clinical adoption is often the case for off-label use of such tests, or for tests that enter the market as a home-brew clinical laboratory test. But premature and inappropriate adoption of biomarker tests could be even costlier for society, the group noted. According to Dr. Ramsey, "if a biomarker diffuses into clinical use and we really do not know what it is doing to folks, the cost of that could be enormous and exceed the cost of doing a clinical trial itself."

Members of the discussion group stressed that clinical trials of biomarker tests should be designed so that the diagnostic test is tied to a therapeutic intervention. "There has to be a plan for what you do if the test is positive and that has to be built into the trial early on," said Dr. Norton. But other group members asserted that conducting such trials early in biomarker development is not feasible because there are not enough resources (both funds and patients) to conduct such large expensive trials for all the biomarkers currently in development.

Discussant Walter Koch, PhD, of Roche Molecular Systems suggested that such trials be reserved for screening biomarker tests for which there is a diagnostic test that can help determine whether the cancer it detects would need to be treated; for example it may be indolent. He also suggested additional criteria for selecting the most promising candidate tests on which to conduct clinical trials; criteria would include a clinical need for the tests and the availability of effective treatments for the cancers they detect. Another criterion suggested by group members was that the potential screening biomarker should first show promising results when tested using well-annotated archived patient specimens collected for other prospective clinical studies, such as the Women's Health Initiative or the NCI's Prostate, Lung, Colon, and Ovarian screening trial. Group members said a conceptual framework was needed for deciding which biomarker tests should proceed to clinical trials.

The group also discussed which type of biomarker tests—those based on multiple markers versus those based on a single marker—are more advantageous to pursue. Discussant Hongyue Dai, PhD, of Rosetta Inpharmatics said he thought multiple markers are more powerful in terms of measurement certainty, sensitivity, or specificity. "With multiple markers, you can do a pattern match, and not simply rely on a high or low judgment based on threshold," he said.

Others were concerned about the use of a pattern biomarker test that stemmed from overfitting of data, but Dr. Dai said overfitting could be avoided by predefining the pattern prior to testing its predictive power. He added that a pattern biomarker test that considers multiple steps on the same cancer-causing pathway is more likely to be accurate than one that relies on just one step of the pathway. But it can be difficult and arbitrary to draw the threshold on pattern biomarker tests as well, according to Brian Taylor, PhD, of Expression Pathology. "Let's say you are looking at 20 different biomarkers [in your pattern test]. If 12 of those fit, does that mean that your test has worked, or does it have to be 8 or 17? How do you draw those lines?"

The group recognized that NIH traditionally has not funded translational work for biomarker tests, so it is difficult to find funding to run clinical trials on them. There is also a lack of incentives for academics to undertake such trials because the academic career and reward structure does not encourage translational work. In addition, few incentives exist for industry to undertake such costly long-term clinical trials, which will not necessarily reward companies with higher revenues, Dr. Ramsey said.

MECHANISMS FOR DEVELOPING AN
EVIDENCE BASE DISCUSSION

Dr. Sawyers was the moderator who summarized his group's discussion on mechanisms for developing an evidence base. He pointed out that many people who participated in the discussion also participated in the earlier discussion on clinical development strategies for biomarker utilization. Consequently, some of his group's conclusions echo those of the earlier group, which was summarized by Dr. Friend.

Members of Dr. Sawyers's group suggested creating public-private consortia to develop different types of biomarkers. Participants in each consortium would be those parties most likely to use and benefit from the type of biomarker the consortium develops. For example, surrogate endpoint markers are beneficial to all parties conducting clinical trials for the purpose of achieving FDA approval for a drug to enter the market. An example of such a consortium was one created to develop CD4 count and HIV viral load as surrogate endpoints for clinical trials used to gain FDA approval of various antiretroviral drugs for HIV infection. Participants in this consortium were the pharmaceutical companies that were developing drugs to treat patients with HIV.

Biomarkers that predict adverse reactions to drugs actually protect the public so their development should be a publicly funded goal, proposed one discussant. But Dr. Sawyers said that pharmaceutical firms should help pay for their development as they help these companies decide whether to subject drugs to further testing in clinical trials. Consortia to develop pathway biomarkers were also suggested by group members, who broke these biomarkers down into two subcategories: signaling biomarkers and cellular response markers.

Signaling markers detect aberrations in a specific biochemical signaling pathway in tumor cells. For example, these markers include disease-causing changes in the ras or the EGFR genes or proteins. Signaling markers are best suited for determining a prognosis and for choosing an appropriate treatment plan. Cellular response markers measure more general processes such as tumor cell proliferation, apoptosis, or angiogenesis. Ideally, these markers should be measured noninvasively, via serum tests or imaging, to reveal whether tumors are progressing and how a treatment is affecting the targeted tumor.

The development of surrogate endpoint markers, adverse reaction biomarkers, and pathway biomarkers would be precompetitive activities that should not require exclusivity. Therefore, all interested parties would benefit by pooling their activities and sharing the development costs, Dr. Sawyer noted. This is in contrast to diagnostics that will be used only when paired to the use of specific drugs, such as the HercepTest, which is used to predict response to Herceptin. The group suggested that both the diagnostic company and the drug company for a paired diagnostic and treatment share the development costs for these types of biomarkers.

After group participants suggested that public-private partnerships could be established to facilitate development of candidate biomarkers, they explored further which groups should be involved in these partnerships. As previous discussions have noted, academia does some discovery work on biomarkers. But academia typically is not involved in the development of robust diagnostic assays because of a lack of expertise in the industrialization aspects and because of a lack of academic rewards and funding sources for this type of research. Start-up diagnostic companies also are not likely to develop biomarker assays because of the low profit margins of diagnostic tests, which make them unattractive to investors. "There was some discussion that if we wait and hope that this happens through free enterprise, we could be waiting awhile," Dr. Sawyers noted. Consequently, group participants suggested a national effort to drive biomarker development, with NCI as the most likely agency to further this effort and support academic researchers doing this type of work.

A public-private partnership that furthers biomarker development could be modeled after the SNP Consortium. This nonprofit foundation was organized to provide public genomic data, and it was supported by pharmaceutical and technical companies and the Wellcome Trust medical research charity. One discussant indicated that a main impetus for forming the foundation was to prevent academic institutions and industry from claiming intellectual property rights on each SNP they discovered in the human genome. Avoiding intellectual property claims could be an impetus for starting a biomarker consortium as well. The group noted that such claims on each possible biomarker could be a huge impediment to having diagnostic companies develop assays for the biomarkers. Several people in the group felt strongly that biomarker information should be in the public domain, with some stating "the real value of the intellectual property comes from developing the assays and not just linking an mRNA to a possible outcome," Dr. Sawyers reported. This raised the problem of how to give diagnostic companies exclusive rights so that they are encouraged to fully develop and commercialize a biomarker.

The group came up with several incentives for biomarker development. Defining the FDA approval path for a biomarker diagnostic more clearly, and linking the approval path for paired diagnostics and therapeutics so both companies share the risks and development costs would provide incentives for biomarker development. It was also suggested that there be patent extensions of innovative biomarker diagnostics to reward the ground-breaking work that one or two companies do that is then used by competing companies to develop similar products. Precedents exist for this enhanced exclusivity in the development of pediatric interventions, and have been proposed for the development of anti-infectives, Dr. Frank noted.

Finally, group participants suggested working with payors to define the cost effectiveness of biomarker tests. "There was a sense that the cost effectiveness of a biomarker was not really appreciated," he said. "If it were, then reimbursement paradigms could be built in that would incentivize companies to make them sooner." Group discussants also suggested working with payors to establish alternatives to basing reimbursement decisions on evidence generated from large, long-term clinical trials. CMS and other insurers often require more evidence than does the FDA for a biomarker's effectiveness prior to reimbursing its clinical use, Dr. Sawyers noted. Several group members suggested that evidence could be generated after the test enters the clinic via community-based postmarketing studies. These studies could be facilitated by using an electronic medical records infrastructure.

Dr. Sawyers concluded his summary by discussing his group's suggestion that there be a demonstration project to develop biomarkers for drugs already on the market. This project could show the value of using biomarkers to identify the group of patients most likely to respond to the drug, or to identify and exclude those likely to have severe adverse reactions to the drug. Such a proof-of-concept experiment could lay out a path for developing biomarkers and could provide lessons about the appropriate business model to follow and regulatory issues to consider. The reason to use approved drugs for the demonstration project is because patients already taking the drugs can be easily accrued into a study, Dr. Sawyers said. One discussant suggested demonstrating the usefulness of biomarkers that indicate the safety of a number of drugs in a class. Another discussant suggested

using the demonstration project to show the value of biomarkers in predicting responsiveness for two or three drugs widely used in oncology.

If a demonstration project had high-impact findings, it could serve as
a catalyst that would spur investment into diagnostic companies and lead
more academic institutions and industry to pursue biomarker discovery
and development, the group pointed out. Several discussants thought some
"success stories" via such a demonstration project would overcome the
inertia that is preventing extensive biomarker development. The science
needed to do such work is already in place, they noted, and what is lacking
is leadership and funding. As an example of a biomarker demonstration
project, Dr. Sawyers mentioned the pilot project already under way that
was previously described by Dr. Woodcock in her presentation. This is a
nonprofit public-private partnership to qualify FDG-PET as a marker for
drug response in non-Hodgkin's lymphoma. Dr. Sawyers' group also reiterated the need for annotated, quality-assured patient samples that are readily
available to further efforts to develop biomarkers.

EVALUATION OF EVIDENCE IN
DECISION MAKING DISCUSSION

Dr. Ramsey was the moderator who provided the summary of the
discussion on evaluation of evidence in decision making. This discussion
group noted that many biomarker-based tests in wide use today were never
thoroughly evaluated for analytic validity, clinical validity, or clinical utility
in relation to standards. Consequently, their value is often unknown. Group
members suggested that this lack of standardized evaluation be eliminated
for new tests because the developmental and clinical costs of these tests are
quite expensive, and costs also can be incurred if tests are used inappropriately and/or cause undue harm to patients.

Some group participants agreed there is a need for more uniform standards for biomarker evaluation. Dr. Ramsey said there is no consistency
regarding standards among organizations and regulatory programs such
as the FDA, CLIA, the College of American Pathologists (CAP), and the
American Society for Clinical Oncology (ASCO). Each organization has its
own set of standards for biomarker tests that are based on different criteria.
There is even variability within these organizations, the group noted. In
a discussion following Dr. Ramsey's summary, Dr. Dai pointed out that
scientific journals also have their own set of standards for biomarkers. For
example, if researchers want to publish gene expression biomarkers, journals

may ask them to compare the biomarkers to what is already available. They may even require that researchers use a specific statistical modeling technique when making such comparisons.

Group members thought the ASCO guidelines for tumor biomarkers for breast or colorectal cancer[21] could serve as a potentially useful model in terms of how one might set standards for evaluating whether biomarkers are ready for clinical use. These guidelines established the appropriate levels of evidence needed for different types of clinical decisions made based on biomarker test results. For example, the highest level of evidence was required for a biomarker assay that would indicate the need to deny specific care, that is, one that indicates drug resistance.

However, there was no group consensus on what standards should be required or recommended for cancer biomarkers. This lack of consensus stemmed, in part, from recognizing that there is no gold standard for many of the new kinds of assays used to detect cancer biomarkers, and the evolving nature of those technologies. This made many in the group reluctant to specify standards. In addition, the group recognized that broad, generalized standards alone are not sufficient; guidelines may also need to be use specific and even target specific.

Because the technologies for genomics and proteomics assays are rapidly evolving, the group noted, standards have to be adaptable to the changes in technology that are continually occurring. There is also such a wide range of uses for biomarkers in the cancer arena that standards for one use, such as a surrogate endpoint in clinical trials, may not be applicable to another use, such as a predictor of patient responsiveness. In addition, the standards for a biomarker that predicts responsiveness to a drug may vary depending on the type of cancer on which it is tested, such as lung or breast. However, basic generalized criteria should be met for all clinical tests including biomarker-based tests, the group members recognized. They agreed with Dr. Ransohoff's assertion in his presentation that the standards of clinical epidemiology still apply to biomarker-based tests.

Working against a common desire to fully evaluate biomarker tests and ensure they meet certain standards is the desire of companies to bring such tests to market as quickly as possible to generate revenues to compensate for development costs. In addition, companies that are developing biomarker tests to be used in combination with specific drugs are often under time pressure to put the drug and the diagnostic on the market at the same time.

---

[21] *http://www.jco.org/cgi/content/full/19/6/1865.*

Because diagnostic development often lags behind drug development for paired diagnostics and therapeutics, shortcuts may be taken in evaluations of the diagnostic, some discussants pointed out.

Because of such financial and time pressures, companies usually seek the fastest and easiest entry into the market, such as CLIA certification for a home-brew laboratory test, rather than a more rigorous evaluation process by the FDA that might require them to conduct clinical studies. Consequently, few biomarker-based tests are designated Class III devices, which require clinical evidence of their effectiveness and safety.

Competition with other companies also prods the makers of biomarker diagnostics to lower the standards bar in order for their products to go to market before those of their competitors. As Sharon Kim, MBA, of Precision Therapeutics observed, "The challenge has been not just to set your own quality standards for yourself, but you worry and wonder what your potential competitors might be held to because there is no standard, and so are you holding yourself to too stringent of a standard, knowing there may be someone else out there that may place a lower level-of-evidence bar or variability bar out there? While the FDA has the ability to come in and regulate, they have elected not to, and so it is more self-regulated. Even for CLIA-governed or CAP-governed labs, there is no specific cookbook or guidance you can go to."

Industry representatives in the discussion group pointed out that companies often evaluate their biomarker diagnostics in phases, with a more complete evaluation of their broader applications not occurring until after the tests enter the market. For example, a cancer detection test may at first only be evaluated for its accuracy and predictive value in high-risk populations because this evaluation can be done relatively quickly compared to one done in the general public. But to create a greater market for their products, companies may evaluate them for broader uses once they are already on the market for a more restrictive indication. In that way, companies can quickly bring their products to market and begin gaining revenue on them to help cover the costs of further evaluations. But once a test is on the market, there are few ways to control, beyond coverage decisions, how the test is clinically used.

As was noted in Mr. Heller's presentation at the conference, the high-variability and rapidly evolving approach to the FDA's regulation of biomarker diagnostics has created uncertainty as to what evaluations industry needs to do of their tests and what standards to apply, Dr. Ramsey said. The group spent some time discussing whether health insurance payors should set the standards for biomarker diagnostics. The group noted that if they did, it would add another layer of variability, uncertainty, and complexity that would be problematic for the developers of the tests, especially if there was no agreement among health insurers in this regard.

The group also considered whether the FDA, CMS, and perhaps other stakeholders should work together to develop more uniform standards for the evaluation of biomarkers. But consensus was not reached on this issue, in part because of the tradeoffs involved. Having these agencies set uniform standards would be beneficial in that companies would know what to expect and what would be required of them regarding the evaluation and performance of their biomarker diagnostics. "As long as they are not overly burdensome, they would help us defend our experimental design if we could refer to something else that had been published and widely accepted. That way when the data were reviewed our study design wouldn't be questioned, which could help speed things through [an FDA approval process]," said Lynne McBride of BD BioSciences.

But Dr. Aronson, the session moderator, said, "there are decisions that come out of CMS and the FDA that are more political than rational and health plans do not follow them." But she added that it would be valuable to gather together a community of stakeholders to help establish the evidence base needed for biomarkers used clinically.

In a discussion following Dr. Ramsey's summary, Dr. Waring stressed the need to engage the pathology community when setting standards for biomarker tests. "When we are talking about predictive tests that determine treatment decisions for patients with serious life-threatening diseases, I think that the pathologists and the pathology community are often the afterthought in this process. We need to engage them very early and make sure they understand the consequences of the decisions and that they maintain quality testing," he said. He noted that CAP and ASCO would be meeting in a few weeks to try to develop common guidelines for HER2 testing.

The group discussed further Dr. Waring's presentation on the variability among laboratories on the accuracy of the IHC test for HER2. Part of that variability stemmed from the manual, visual, and subjective nature of the test, the group noted. But it is likely that such variability in accuracy will crop up again for other biomarker tests, Dr. Ramsey said. The group debated whether there should be additional measures of quality assurance in such tests. Suggested quality assurance measures included proficiency testing akin to what is now required for cytotechnologists who read Pap smears, volume requirements akin to what is required for radiologists who read mammograms, and requirements for collecting, analyzing, and reporting data on test performance. There was no consensus on which measures, if any, should be pursued to improve the quality of biomarker testing.

INCORPORATING BIOMARKER EVIDENCE INTO CLINICAL PRACTICE DISCUSSION

Moderator Robert McDonough, MD, of Aetna U.S. Healthcare summarized his group's discussion on incorporating biomarker evidence into clinical practice. He noted that there are many sources of information on biomarkers that reach clinicians, including journals, colleagues, product vendors, patients, popular media, practice guidelines, clinical trials abstracts, meetings, and continuing medical education. But when the group evaluated what prompts clinicians to adopt biomarker tests into their clinical practices, evidence-based information was not high on the list. "If you are looking at the screening for cancers, there is no correlation between the strength of the evidence and adoption," said discussant Mark Fendrick, MD, of the University of Michigan.

For example, an impressive 75 percent of the target population undergoes regular screening for prostate cancer, despite the fact the USPSTF gave it an unimpressive I rating. This is in contrast to the 50 percent of the target population who undergo regular colon cancer screening, which the USPSTF gave its highest rating because of its proven effectiveness. Academic practitioners appear to be more influenced by evidence, however, and may delay adopting a new test until there is evidence showing its effectiveness, several discussants agreed. This is in contrast to community practitioners, who may more readily adopt a new test or drug, even when there is little to no evidence of its clinical value. As a consequence, once a product enters the market, it may be impossible to gather the evidence on a test's clinical value because of difficulties accruing patients to serve as controls for the trials needed to gather that evidence.

Other factors beyond evidence appeared to be more important in influencing the incorporation of biomarker tests into clinical practice, the group noted. The most influential factor they identified was reimbursement for a test at a sufficient level. "If you look at the adoption of CT scans, PSA testing, or even COX-2 inhibitors, until they were paid for, they were not used," said Dr. Fendrick. Because most diagnostics are relatively inexpensive, insurers are more likely to reimburse their costs without scrutinizing the evidence base for the test, the group also noted. "If they didn't pay for even low-cost biomarkers unless they were validated in a proper way, that would be an incentive to do those [validation] studies," said discussant Dr. Carbone.

The promotion that health insurers and employers do for various tests also influences their use, some discussants pointed out. For example, insurers often promote preventive health tests, such as those used to screen for various cancers, via informational mailings and their websites. "Some employers give discounts on health insurance to employees who undergo a self-assessment that indicates what types of screening and other health maintenance measures they should undertake," Dr. Carbone said. "I think it is widely adopted when you give people a buck to do it."

Another highly influential factor was whether the test was adopted by what the group called "thought leaders." A thought leader is someone who other members of a group look to as an authority. A thought leader may be misinformed, but he or she is still influential. In academic settings, thought leaders tend to be the lead investigators of clinical studies or the chairs of departments. In clinical practices the thought leader "is the clinician down the hall who seems to be knowledgeable about what is new in medical technology," Dr. McDonough said. He said one discussant noted that physicians who practice in groups seem to adopt technology more rapidly than solo practitioners, possibly because of the presence of thought leaders in group practices.

Another potential driver for the uptake of new biomarker tests is patient requests for the tests, the group noted. Studies reveal that if a patient asks for a drug by name, there is an 80 percent chance that a physician will prescribe it, Dr. Fendrick observed. Presumably patients have the same influence over the tests they request, he suggested.

Through promotional efforts, product manufacturers also influence doctors and patients to use their biomarker tests, Dr. McDonough noted. "What I always thought was an important factor was the guy who knocks on your door—the vendor of the new device or new drug or new test," he said during his group's discussion. Dr. Waring also noted that for a test such as the FISH test for HER2, used to determine patient responsiveness to a specific treatment, the pharmaceutical company that provides that treatment may pay the costs of the test if it is not covered by an insurance provider. This is especially the case in Europe where national health plans may not offer the test as part of their services. "Roche until recently was paying for those tests to be performed in their own central laboratories," he said. "So these tests were becoming available not because of reimbursement issues—they were being made available by the pharmaceutical company for business reasons."

Other influences on the clinical adoption of a biomarker test hinge on features of the test itself, the discussion group said. Ease of interpretation is one such feature. If the test is easy to interpret and has a simple positive versus negative result, it will be adopted more readily than a test whose results require "some kind of complex algorithm to understand," said Dr. McDonough. Clinicians are also more inclined to adopt tests that are reliably accurate and have timely results. "If you need to make a decision today, and the test is going to take 2 weeks, regardless of how easy or reliable that test is, it may not be very clinically useful," said Dr. McDonough. Clinicians are also more likely to adopt tests if there is little to no risk in using them, and there are no alternative tests or test-linked treatments. Insurers are also more likely to reimburse for both the test and treatment, for those that are linked, if there are no treatment alternatives and the disease the drug targets is life threatening, the group noted.

Inconvenience to the patient is another important test feature that influences its adoption in the clinic. Physicians are more likely to prescribe a simple blood test than an endoscopic procedure or a test that requires a stool sample, Dr. McDonough pointed out. Practitioners are also more likely to use a test that will influence their clinical decision making. "Is it a test that might give you some idea of the prognosis of lung cancer, but will not actually influence the type of therapy you might actually give to the patient? If the test does not seem to have any influence on the clinical management we would hope that would make it less likely that a clinician would use it," Dr. McDonough said.

Like other discussion groups, Dr. McDonough's group recognized that low profit margins on diagnostic tests act as a disincentive to the development of biomarker tests and their evaluation in clinical trials. This led to the suggestion by Dr. McGivney that payors help subsidize some of this clinical research. "A payor who is asking for evidence should actually support, in part, the development of some of that evidence," he said. Dr. McDonough said that some insurers, such as Aetna, do pay for routine costs of their patients in clinical trials. But Dr. McGivney countered that there is an increasing trend for payors not to cover such costs.

Given that reimbursement levels highly influence the adoption of clinical tests, other discussants suggested that payors tailor their copay amounts for biomarker tests based on a test's value or degree of evidence to support any positive impact on patient outcomes. Zero copayment amounts could be allotted for those biomarker tests that are highly cost effective and likely to affect clinical management. High copayments could be required for tests whose cost effectiveness is questionable due to a lack of evidence on their benefits.

But the group recognized that "it would not be easy to structure a benefit program to that fine a degree of assigning copays based on someone's assessment of cost effectiveness," Dr. McDonough said. There would be legal issues that might be difficult to overcome, such as varying state regulations that affect copayment levels. In addition, both legislators and the insurance clientele might look askance at plans that specify high copayments for treatment-linked tests for life-threatening illnesses.

For payors to more adequately influence the adoption of biomarker tests, those tests need to have their own Current Procedural Terminology (CPT) codes, group members noted. These identifying codes are established by the American Medical Association and are used to report medical procedures and services to health insurers. Health insurers then specify reimbursement rates for each code. CPT codes are also used for developing guidelines for medical care review. "Many of these biomarkers do not have specific CPT codes," said Dr. McDonough. "They are defined by process steps so that the insurer, even if they were willing to scrutinize biomarkers, often find it difficult to know what type of biomarkers are being used. What this means is that many of these biomarkers are being incorporated into clinical practice without much scrutiny."

This is especially true for home-brew tests, which are always defined by process steps. These tests, therefore, bypass scrutiny by both regulators and reimbursers, the group noted. Even when a test has been approved by the FDA, some discussants said, there is no guarantee that laboratories will use that test. Instead, they may offer their own home-brew version of the test, which may not be as acurate. Home-brew versions of the HercepTest, Dr. Waring said, help explain the variability in accuracy among laboratories.

In a discussion following Dr. McDonough's summary, Dr. Ramsey gave an overseas perspective of health care payors playing a role in gathering clinical data to evaluate new products. For example, the United Kingdom's National Health Service pays for a new drug at an agreed upon price, with the requirement that data on the drug's effectiveness be collected in a patient registry. If the drug does not show effectiveness at the expected level, the drug's price is reduced so that the total reimbursement over time reflects the actual quality of life gain observed. He thought such risk sharing in drug development was valuable, and noted that the group's suggestion that payors cover the costs of clinical trials on biomarker tests would put all the burden of risk on insurance companies. He suspected they would balk at such a suggestion and reiterated that risk sharing has some value.

REFERENCES

Check E. 2004. Proteomics and cancer: Running before we can walk? *Nature* (429), 6991: 496-497.

Ellis IO, et al. 2004. Best Practice No 176: Updated recommendations for HER2 testing in the UK. *Journal of Clinical Pathology* 57:233-237.

FDA (Food and Drug Administration), Center for Devices and Radiological Health, OVID (Office of In-Vitro Devices), Analyte Specific Reagents; Small Entity Compliance Guidance; Guidance for Industry.

Letter from OIVD to Roche Molecular Diagnostics Re: AmpliChip, *http://www.fda.gov/cdrh/oivd/amplichip.html.*

Medical Devices; Classification/Reclassification; Restricted Devices; Analyte Specific Reagents, 61 Fed. Reg. at 10,484. March 14, 1996.

Michiels S, et al. 2005. Prediction of cancer outcome with microarrays: a multiple random validation strategy. *Lancet* 365(9458):488-492.

Paik S, et al. 2002. Real-world performance of HER2 testing—National Surgical Adjuvant Breast and Bowel Project experience. *Journal of the National Cancer Instititue* 94(11):852-854.

Perez EA, et al. 2006. HER2 testing by local, central, and reference laboratories in specimens from the North Central Cancer Treatment Group N9831 intergroup adjuvant trial. *Journal of Clinical Oncology* 24(19):3032-3038.

Petricoin EF, et al. 2002. Use of proteomic patterns in serum to identify ovarian cancer. *Lancet* 359(9306):572-577.

Pharmaceutical Research and Manufacturers of America Biomarker Working Group presentation (2004). FDA Advisory Committee Meeting.

Reddy JC, et al. 2006. Concordance between central and local laboratory HER2 testing from a community-based clinical study. *Clinical Breast Cancer* 7(2):153-157.

Rhodes A, et al. 2004. The use of cell line standards to reduce HER-2/neu assay variation in multiple European cancer centers. *American Journal of Clinical Pathology* 122:51-60.

Simon R, et al. 2003. Pitfalls in the use of DNA microarray data for diagnostic and prognostic classification. *Journal of the National Cancer Institute* 95(1):14-18.

Van de Vijver MJ, et al. 2002. A gene-expression signature as a predictor of survival in breast cancer. *New England Journal of Medicine* 347(25):1999-2009.

Wagner JA. 2002. Overview of biomarkers and surrogate endpoints in drug development. *Disease Markers* 18(2):41-46.

WORKSHOP AGENDA

National Cancer Policy Forum
Workshop on
Developing Biomarker-Based Tools for Cancer Screening,
Diagnosis, and Treatment:
The State of the Science, Evaluation, Implementation, and Economics

National Academy of Sciences Building Auditorium
2101 Constitution Avenue, N.W.
Washington, DC

Agenda
2.5 days, March 20-22, 2006

Day 1—March 20, 2006

8:30 am Welcome and introductory remarks
Hal Moses, MD (Vanderbilt University, Chair, National Cancer Policy Forum)

8:45-10:15 am Session 1
Brief overview of technologies
Moderator: Howard Schulman

Presentations:

Genomics-based technologies (including DNA microarrays, CGH, and sequencing technologies)
    Todd Golub, MD (The Broad Institute of Harvard and MIT)

Proteomics and metabolomics technologies
    Howard Schulman, PhD (PPD Biomarker Discovery Sciences)

Technologies for physiological characterization (including functional imaging)
    Michael Phelps, PhD (University of California, Los Angeles)

10:30 am-12:00 noon Session II
Overcoming the technical obstacles
Moderator: Charles Sawyers

Presentations:

Informatics and data standards
John Quackenbush, PhD (Harvard)

Biomarker validation
David Ransohof, MD (University of North Carolina)

Biomarker qualification: Fitness for use
John Wagner, MD, PhD (Merck and Co., Inc.)

12:00 noon-1:00 pm Lunch break

1:00-3:00 pm Session III
Coordinating the development of biomarkers and targeted therapies
Moderator: David Parkinson

Presentations:

Therapeutics industry perspective/realities (examples of successes and difficulties/failures of targeted therapy)
Paul Waring, PhD (Genentech)

Diagnostics industry perspective (industry mission/business models/marketing strategies, & IP)
Robert Lipshutz, PhD (Affymetrix)

NCI/NIH perspective (goals and funding initiatives)
James Doroshow, MD (National Cancer Institute)

Clinical investigator perspective
Charles Sawyers, MD (University of California, Los Angeles)

3:15-5:45 pm Small Group Discussions
Policy implications surrounding biomarker development—
prioritizing problems and solutions

1) Strategies for implementing standardized biorepositories
Moderators—Carolyn Compton, Brent Zanke, Hal Moses
Invited Discussants—Edith Perez, Margaret Spitz, B. Melina Cimler,
Indra Poola, Ann Zauber

2) Strategies for determining analytic validity and clinical utility of
biomarkers
Moderators—Janet Woodcock, Howard Schulman, John Wagner
Invited Discussants—Walter Koch, Zoltan Szallasi, Scott Patterson,
Ronald Hendrickson, David Carbone, Laura Reid

3) Clinical development strategies for biomarker utilization
Moderators—Charles Sawyers, Stephen Friend, David Parkinson,
Richard Simon
Invited Discussants—Richard Schilsky, David Agus, Barbara Weber,
Richard Frank, Robert Gillies

4) Strategies to develop biomarkers for early detection
Moderators—Scott Ramsey, David Ransohof
Invited Discussants—Jean-Pierre Wery, Kathryn Phillips, Larry
Norton, Hongyue Dai, David Muddiman

5:45 pm Adjourn Day 1

Day 2—March 21, 2006

8:30 am Welcome—Opening remarks
Hal Moses

8:45-10:15 am Session IV
Biomarker development and regulatory oversight
Moderator: Janet Woodcock

Presentations:

FDA Critical Path Initiative
    Janet Woodcock, MD (Food and Drug Administration)

Clinical laboratory diagnostic tests: Oversight for analytical and clinical validation
    Mark Heller, JD (Wilmer Cutler Pickering Hale and Dorr)

Clinical trial design and biomarker-based tumor classification systems
    Richard Simon, DSc (National Cancer Institute)

10:30 am-12:00 noon Session V
Adoption of biomarker-based technologies
Moderator: Alfred Berg

Presentations:

CMS strategies for biomarker coverage
    Jim Rollins, MD, PhD (Centers for Medicare & Medicaid Services)

Insurance coverage and practice guidelines
    William McGivney, PhD (National Comprehensive Cancer Network)

Technolgy assessment and clinical decision making
    Alfred Berg, MD, MPH (University of Washington)

12:00 noon-1:00 pm Lunch Break

1:00-2:30 pm Session VI
Economic impact of biomarker technologies
Moderator: Scott Ramsey

Presentations:

Cost-effectiveness analysis and technology adoption in the UK
  Andrew Stevens, MD (UK National Institute for Health and Clinical
  Excellence)

Cost-effectiveness analysis and the value of research
  David Meltzer, MD, PhD (University of Chicago)

The payer perspective
  Naomi Aronson, PhD (BlueCross BlueShield Technology Evaluation
  Center)

2:45-5:15 pm Small Group Discussions
Policy implications surrounding biomarker adoption—prioritizing
problems and solutions

1) Mechanisms for developing an evidence base
  Moderators—Janet Woodcock, David Parkinson, Charles Sawyers
  Invited Discussants—Walter Koch, Indra Poola, Laura Reid, Richard
  Frank

2) Evaluation of evidence in decision making
  Moderators—Naomi Aronson, Scott Ramsey
  Invited Discussants—Ronald Hendrickson, Ann Zauber, Kathryn
  Phillips, Barbara Weber, Robert Gillies

3) Incorporating biomarker evidence into clinical practice
  Moderators—Robert McDonough, William McGivney
  Invited Discussants—David Carbone, David Agus, Hongyue Dai,
  Mark Fendrick, Judith Hellerstein, Judith Wagner

5:15 pm Adjourn Day 2

Day 3—March 22, 2006

Reports from small group discussions

8:30-10:00 am Reports from day 1 group leaders

10:15 am-12:15 pm Reports from day 2 group leaders

12:15 pm Wrap up/summary
Hal Moses

12:30 pm Lunch—Adjourn